Living Off the Land in Space

LIVING OFF THE LAND IN SPACE

Green Roads to the Cosmos

Gregory L. Matloff
Les Johnson
C Bangs

Copernicus Books
An Imprint of Springer Science+Business Media

In Association with
Praxis Publishing Ltd

Published in the United States by Copernicus Books,
an imprint of Springer Science+Business Media.

Copernicus Books
Springer Science+Business Media
233 Spring Street
New York, NY 10013
www.springer.com

Printed on acid-free paper.

9 8 7 6 5 4 3 2 1

ISBN 978-1-4899-8877-5   ISBN 978-0-387-68316-4 (eBook)

CONTENTS

Contents

Contents

Contents

FOREWORD

Two scientists and an artist have collaborated to produce this book, which is both scientifically authoritative and artistically inspiring. Les Johnson, a physicist by education, managed in-space propulsion research for about a decade at NASA's Marshall Space Flight Center in Huntsville, Alabama, and is now the manager of that Center's space science programs and projects office. Dr Gregory Matloff, a member of the International Academy of Astronautics and a Fellow of the British Interplanetary Society, has published many technical papers and two books on these topics and has consulted for NASA. C Bangs has shown her space-related art in many international settings; her work enlightens the visionary drive in humanity towards cosmic exploration.

As well as qualifying as an instruction manual for scientists and engineers, this book should appeal to far-sighted minds who feel that the new millennium will hopefully see the expansion of humanity beyond the edges of the solar system. It is hoped that this book will inspire astronomical researchers currently seeking Earth-like worlds circling nearby stars. In centuries to come, these worlds may well serve as new homes for humanity.

Although not science fiction, this is a visionary book. Already, researchers in the SETI (Search for Extraterrestrial Intelligence) community have greatly expanded their search for alien radio transmissions. The time may well come when deep-space exploration and SETI will join forces to permit humanity to join a wider galactic community.

Dr Claudio Maccone
Co-Vice-Chair, SETI Permanent Study Group,
International Academy of Astronautics

ACKNOWLEDGMENTS

Many people contributed to the research described in this book. The cited technical and popular references outline their efforts.

Artist C Bangs collaged digital photographs, created and scanned her drawings and internet imagery to produce cover and chapter frontispiece art. All planetary and spacecraft images are from NASA websites; all star fields and galactic images are from the Hubble Space Telescope websites.

The image of the Conestoga Wagon collaged into the Chapter 1 frontispiece is used with the courtesy of the Detroit Michigan Historical Museum.

A photograph of a steam locomotive is incorporated in the Chapter 12 frontispiece. We thank Dipl.-Ing. Tobias B. Kohler of Graz, Austria, for his permission to use this image.

The frontispieces for Chapters 4 and 11 are derived from an aerocapture poster produced by the In-Space Propulsion Research Group at NASA Marshall Space Flight Center. C Bangs participated in the creation of this poster during her summer 2004 tenure as a NASA Faculty Fellow.

During the summer of 2003, C Bangs participated in the creation of another In-Space Propulsion Research Group poster project at NASA Marshall that combines elements of ancient planetary mythology and modern space technology. This poster is used as the frontispiece for Chapter 21.

In 2001, C Bangs was funded by NASA Marshall to create the prototype Rainbow Holographic interstellar message plaque described in Chapter 14. Some of the six multiplexed two- and three-dimensional images on that plaque are incorporated in that chapter's frontispiece.

The poetry and folksong segments used to introduce each chapter come from several sources. These include Louis Untermeyer, *A Treasury of Great Poems*, Simon & Schuster, New York (1942) and Walt Whitman, *Leaves of*

Grass, ed. E. Holloway, International Collectors Library, Garden City, New York (Doubleday copyright 1926)

We are also grateful to our colleague Dr Claudio Maccone for consenting to write the Foreword to this volume. The efforts of our publisher, Clive Horwood, are also appreciated, as are the contributions of the editorial staff of Praxis and Copernicus Publishing Companies. Special thanks go to Praxis editors Arthur Foulser and Alex Whyte and to Harry Blom and Christopher Coughlin of the New York Springer office.

Discussions with Paul Farrell and Paul Gilster are also gratefully acknowledged. The authors have cross-checked the draft manuscript many times in an effort to minimize errors and we take full responsibility for any errors that may have escaped this process.

INTRODUCTION

CITIZENS OF THE COSMOS

Full fathom five thy father lies:;
Of his bones are coral made;
Those are pearls that were his eyes;
Nothing of him that doth fade
But doth suffer a sea-change
Into something rich and strange.

William Shakespeare, from *The Tempest*

A PERSON younger than 40 or so can be forgiven for imagining that space travel has always been with us. Those a bit older, who matured in the heady days of humanity's first hesitant steps into the cosmic abyss, are fortunate enough to have witnessed the drama.

In less than half a century we have collectively altered. No longer are we a two-dimensional planet-dwelling folk, forever bound by the vagaries of soil, ocean, wind, and rain. We have climbed above the clouds, tasted the vacuum of low Earth orbit (LEO). Some of us have lived above the atmosphere for a year or longer; a few dozen have orbited our planet's solitary natural satellite or disturbed its dusty surface with their bootprints. On their fiery return to Earth, the pioneering astronauts of these Apollo expeditions returned rocky samples of the lunar surface to terrestrial laboratories.

Our robotic emissaries, less concerned with the requirements of life support, have ventured further afield. They have flown by, circled, or landed upon every solar-system planet save tiny, frigid Pluto (which is not actually a planet). Probes from Earth have touched down upon hellish Venus, roved the frozen deserts of Mars in searches for life and water, entered the atmosphere of giant Jupiter, and successfully reached the surface of Saturn's giant moon, Titan. Many other small solar-system bodies—asteroids, comets, and planetary satellites—have been explored by these small ships from Earth. Carrying engraved messages, four of them have actually departed from the solar system to endlessly cruise the galactic void as the first starships from Terra. Stationed above our planet's turbulent atmosphere, sophisticated space telescopes sensitive to many regions of the electromagnetic spectrum have extended our intellectual reach by billions of light-years, and searched billions of years into the past.

But as well as exploring, astronauts and robots have begun to lay the framework and infrastructure for an extraterrestrial economy. Already, most long-distance communication is routed through geosynchronous communication satellites permanently stationed about 35,000 kilometers above the equator. Other spacecraft monitor weather and climate and help to record our planet's extensive, but finite resources. It's difficult to get lost anywhere on our planet, when one's location can be easily ascertained by routinely tapping into the global-position satellite network.

Perhaps our dreams of cosmic flight began with observations of birds and other living fliers. Although there are legends of ancient people experimenting with kites and hang gliders, the first successful device to carry humans above the Earth's surface was the hot-air balloon. Only a century or so after the Montgolfier brothers' first balloon ascent above the French countryside, other inventors such as the German Count Zeppelin learned how to put reciprocating engines on board lighter-than-air aircraft and control their flight through the sky.

But although these early sky ships were cumbersome and slow, the hydrogen gas that replaced hot air had a nasty tendency to react explosively with atmospheric oxygen.

In the first quarter of the twentieth century, heavier-than-air aircraft began to replace the lumbering giants. Initially driven by propellers, early airplanes proved to be safer, faster and more economical than the giant Zeppelins.

The Second World War was a watershed in the development of powered flight both within and above the atmosphere. The jet engine was developed as a replacement to the propeller; and both the speed and cruising altitude of commercial and military aircraft increased as a consequence of this development. Before 1950, aircraft had exceeded Mach 1, the sound barrier, and achieved operational altitudes in the lower stratosphere.

Another reaction motor, the rocket, saw its first large-scale application during the Second World War. Some of these rocket-propelled ballistic missiles crossed the 60-kilometer threshold to space as they arched toward their distant targets.

The computer also emerged as a computational tool in that period, with some crude automatic devices being used to control the trajectories of early cruise missiles.

At the end of that conflict, the energy of the atom was unleashed destructively. The Faustian bargain of nuclear energy may still prove to be a *blessing* by unlocking vast energy reserves for human applications, or as a *curse*, as it may yet destroy us.

After the conclusion of the Second World War, the victorious powers realized the potential of combining the rocket, computer, and nuclear warhead. Conceptually, a general could arrive at work in the morning and, before his morning coffee break, could push a button and launch a fleet of nuclear-tipped, computer-guided ballistic missiles towards an opposing country. By lunchtime, he could have caused the death of one hundred million people.

The Space Race of the 1960s was neither a quest for knowledge nor an attempt to expand the frontiers of terrestrial life. Plainly and simply, it was an effort by the superpowers—the USA and the USSR—to gain the high ground militarily and impress the developing world with their techno-logical superiority.

The USSR jumped to an early lead in the Space Race with multiple accomplishments, including the first orbital spacecraft (Sputnik 1, in 1957) and the first human-occupied spacecraft (Vostok 1, which carried Yuri Gagarin into orbit in 1961).

Playing catch-up and hoping to regain technological supremacy, America aimed for the Moon, which orbits the Earth at an average distance of 384,000 km. Apollo 8 orbited the Moon in 1968, and Apollo 11 followed with the first piloted Moon landing in 1969.

Although subsequent Apollo missions, especially the final three flights (Apollos 15–17), would return much scientific data about our planet's natural satellite, many people lost interest in the space program after the success of the first lunar expeditions. Clearly, the quest for national prestige had fueled the early Space Age, not the search for scientific data or the desire to open new frontiers.

Since 1972, no human has ventured more than a few hundred kilometers from Earth's surface. Several hundred humans have experienced the weightless environment on board the reusable American space shuttle, Russian Salyut, and Mir space stations and, most recently, the international space station (ISS).

The space-launch fraternity is also expanding. China has demonstrated a capability to launch humans to LEO that rivals that of the USA and Russia. Heavy- and medium-lift launch vehicles are in routine operation in Europe, India, and Japan. Many other nations will soon have the capability to orbit small or mid-sized satellites. And the X-Prize competition demonstrated that suborbital space, at least, is not off limits to private astronauts.

With so many players in this cosmic game, humanity seems poised to expand its reach once again beyond LEO. But what are the motivations that could trigger this expansion when it begins?

The very human desire to soar like the birds, which motivated early aviation pioneers, does not seem applicable in the case of expanding into a vacuum far above the reach of any avian. In an era of one super power, the desire for enhanced national prestige also seems to be an inappropriate trigger for human cosmic expansion.

But business now has a global reach. In the quest to make a profit, private "space liners" will soon apply the technology of Rutan's Space Ship 1 to allow space tourists a few minutes of weightlessness and a glimpse of black skies above the distant Earth for a cost of "only" a few hundred thousand dollars. Some day, as technology improves, these costs will come down and typical citizens of the developed world will easily acquire astronaut wings. Ultimately, this quest for profit may even encourage the development of orbital hotels and the reusable, single-stage-to-orbit ferries they will require. But because of the huge capital costs and time scales involved, few venture capitalists will ante up the huge sums required to begin the economic development of the solar system.

What may be needed is a collaboration between government and private interests. Enlightened governments' interests in solar-system settlement might be two-fold: obtaining the resources of the solar system and insuring national survival.

Less than one-billionth of the light emitted by our Sun actually strikes the Earth. One way of maintaining our planet's high-energy lifestyle in the post-fossil-fuel era is to construct large solar energy collectors in space, using space resources. Using microwave technology, this energy could be beamed back to receivers on Earth's surface.

At approximate intervals of a century, cosmic objects capable of destroying a city impact the Earth. A recent impact in Tunguska, Siberia, in 1908 resulted in an explosion with the energy equivalent of a 20-megaton hydrogen bomb. That's more than enough energy to level a large city and kill 10 million people! Fortunately, the Tunguska object impacted a wilderness.

About 65 million years ago, a much larger space object struck in what is now the Yucatan, Mexico. Most terrestrial species, including the dinosaurs, were wiped out or adversely affected in this event.

Enlightened governments, with an interest in self-preservation, could direct their astronomical resources to accurately track the trajectories of near-Earth objects (NEOs) that might some day threaten the Earth. A task for the military establishments of a consortium of such nations might be the development and maintenance of techniques that could divert Earth impactors.

And while the military is diverting threatening asteroid and comet fragments, entrepreneurs could mine these objects for useful materials, in cooperation with the governmental space programs. The material resources of NEOs could be used to construct solar-power stations to supply terrestrial energy needs and large space habitats to house the space workers (and their families) who are engaged in NEO diversion and mining, and the construction of solar-power stations.

There are terrestrial precedents for such private and public cooperation on the frontier. The settlement of the North American west by private individuals and corporations, for example, was greatly hastened by the construction of transcontinental railroads using public funds.

The breakout into the solar system will be grand and majestic. Unlike previous territorial expansions, it will be a truly international endeavor. Fortunes will be made on this new frontier and lives will be lost; but, from the perspective of the far future, the most important result will be the expansion of the terrestrial biosphere into the celestial realm. Ultimately, Gaia's children will inherit the sky, altered by the new environment to become true citizens of the cosmos.

FURTHER READING

For an exciting journalistic treatment of the Space Race, consult J. Barbour, *Footprints on the Moon* (American Book-Stratford Press, 1969). Early robotic exploration of the outer solar system is reviewed by M.

Washburn, in *Distant Encounters* (Harcourt, Brace, Jovanovich, New York, 1983). In the *Soviet Manned Space Programme* (Salamander, New York, 1988), Philip Clark reviews the Russian space stations through the development of Mir. Prospects for space solar power and asteroid mining are reviewed by John S. Lewis, in *Mining The Sky* (Addison Wesley, New York, 1996).

1

THE OLD FRONTIER

I will arise and go now, and go to Innisfree
And a small cabin build there, of clay and wattles made;
Nine bean rows will I have there, a hive for the honey bee,
    And live alone in the bee-loud glade.

And I shall have some peace there, for peace comes dropping slow,
Dropping from the veils of the morning to where the cricket sings;
There midnight's all a glimmer, and noon a purple glow,
    And evening full of the linnet's wings.

William Butler Yeats, from *The Lake Isle of Innisfree*

THROUGH their million-year history, humans and their hominid progenitor species have been anything but static. Perhaps because of a desire to escape their fellows, perhaps because of a desire for peace, or perhaps for reasons we will never know, our earliest ancestors migrated from an "Eden" in Central Africa to populate Eurasia and Australia. More recently, perhaps 13,000 years ago, the remote descendents of these early pioneers crossed an Ice Age land bridge between Siberia and Alaska to begin the human occupation of the New World.

9

All pre-human and most human territorial expansions occurred before the development of written record-keeping, so it is difficult today to estimate the ratio of successful to failed expansions. But analogies can still be drawn from the archeological record.

PALEOLITHIC MIGRATIONS

Approximately one million years ago during the Old Stone Age (Paleolithic era), bands of *Homo erectus*, the first hominids capable of long-distance bipedal locomotion, trekked across arid terrain connecting Africa and Asia. It may have been population pressure, environmental change, or some unknown factor that sparked the migration of these first, pre-human pioneers.

No longer could these migrants depend upon the mild climate and abundant food supplies of tropical Africa. It was necessary for them to learn various skills enabling survival in unfamiliar environments.

As they climbed the slopes of various mountain ranges, our ancestors would have noticed a distinct decrease in ambient temperature. Even their ample supply of body hair was not enough. Some genius rose to the occasion and noted that skins of slaughtered animals could be used as heat-retaining body covers. Another nameless protohuman observed that natural fires could be maintained by adding wood and leaves and used to cook food and provide heat.

For the success of these early migrations, it was necessary to use the skins of local animals as clothing and to utilize local resources to maintain fires. *Homo erectus* would not have gotten very far if he had to return to Olduvai Gorge in Kenya every time a fire ran down or an animal skin wore out.

The out-of-Africa territorial expansion of *Homo erectus* bands was likely limited by the extent of the Eurasian landmass. It was an early human of our species, *Homo sapiens*, who discovered a method of expanding beyond the Asian continent.

If you consult a map of present-day southeast Asia, hundreds of kilometers separate Australia from major islands of the Indonesian archipelago. But at various geological times sea levels were lower and the water barrier between Aasia and Australia was less daunting.

About 60,000 years ago, some brilliant human must have pondered this reduced waterway and noted that logs and natural rafts could survive ocean voyages. Trees could be cut and joined together with vines to form natural rafts. In periods of calm seas and clear skies, people could float (or perhaps row) across the narrow seas. In this way, the ancestors of the Aboriginal People spread their culture to Australia and New Guinea.

No historians or scientists witnessed this historic migration, but it could not have succeeded unless the migrants quickly learned to use local trees and vines to prepare their island-hopping rafts.

The final Paleolithic migration, and one of the most significant, was the peopling of the New World. Earlier than 13,000 years ago, small bands of humans crossed an Ice Age land bridge from Siberia to Alaska. Perhaps supplemented by a few Europeans who had migrated westward across the Arctic ice sheet, the descendants of these migrants fully occupied North and South America within a few millennia of their arrival. As sixteenth-century European explorers and settlers were to learn, the Native Americans developed cultures that were superbly adapted to the environments they encountered.

NEOLITHIC AND BRONZE AGE MIGRANTS

Around 6,000 BC, a new era of human development, The New Stone Age or Neolithic begins. Human capabilities altered as people began to replace a free-roving, hunter–gatherer existence with settled town life. Architecture improved, as did animal husbandry and agriculture.

In the Mediterranean basin, keen observers must have noted how large nautical birds, such as swans, can utilize their feathers as sails and be pushed by the wind against a river's current. The Nile river boats, some of which have been preserved in tombs, would be instrumental in uniting Egypt into a national entity.

Before the invention of writing, some of these crude craft were used to explore the islands closest to Egypt—Crete and the other Cyclodic isles. Perhaps using craft not unlike those described in the biblical legend of Noah and the Sumerian epic of Gilgemesh, early civilized humans began to island hop across the Mediterranean.

As the Bronze Age dawned with advances in literacy and mathematics, some of the colonists from North Africa and West Asia began to contribute to the art of ship design. Long before the Classical era, keel-equipped ships, originally invented in Minoan Crete, completed the exploration of the Mediterranean Sea. Certainly before 1,000 BC, humans had crossed the English Channel and ventured into the Atlantic.

Once again, native materials must have been extensively used by these explorers and settlers. Their expansion would have been greatly limited if a ship had to return to Tyre, Memphis, Ur, or Knossos every time a sail required mending.

Although the keel was a major innovation, it would provide less-than-adequate stability for an early sailing ship attempting to traverse the stormy Pacific. Several thousand years ago, a genius in New Zealand must have realized that greater stability in rough seas would result if several canoes were lashed together side-by-side to produce the first catamaran.

Over the course of several millennia, the pre-literate Polynesian people used these craft to island hop across the vast Pacific Ocean, sometimes navigating between tiny islands separated by thousands of miles of open sea. Navigation instructions for these epic journeys were passed on in the form of memorized epic poems. Everywhere they ventured, the Polynesians learned to develop and exploit the local environments in their island habitats.

HISTORICAL MIGRATIONS

During the Iron Age, in the fifth century BC, Greek scholars including Thucydides and Herodotus produced the first written histories. Accurate record-keeping subsequent to this development resulted in better knowledge of human territorial expansion during and after the Classical era.

This was the time of the first great empires—the Athenian, Persian, Hellenistic, and Roman. Colonies of existing city states were no longer established only by private individuals; government also played a role.

One model for historical human territorial expansion was established around 900 BC. Recovering from the Aegean Dark Age, Dorian and Ionian Greeks began to expand from their homelands into southern Italy and Asia Minor. What resulted was a "city-hopping" type of expansion in which migrants would first establish a city state mirroring the ideals of their distant home and then, in turn, send out expeditions to establish new colonies. Naples was established in this manner, as a colony of Cumae around 600 BC. Many Greek-established towns around the Black Sea evolved to become major cities of Russia and neighboring countries.

Human population was increasing in this period, and technology was advancing. Unfortunately, this technological advance manifested itself in warfare. Colonies of rival empires engaged in constant battle, in some cases for centuries.

People living in the modern west are used to thinking of progress as a constant upward march, but this is not an accurate historical perspective.

Attempted territorial expansions have not always succeeded. Because of advances in written record-keeping, we know a great deal about historical expansions that failed, as well as those that succeeded.

One notable failure was the expansion of Norse culture starting around 800 AD. The Vikings were a warlike people who preyed upon other cultures they encountered. Viking longboats, equipped with flexible keels, oars and sail, were superbly adapted to the treacherous North Atlantic.

By 930, the Vikings had spread far from their Norwegian origin, to occupy Iceland. In 986, they reached Greenland. Around 1000 AD, it is thought that Viking ships had sailed to Baffin Island, Labrador and to their New World colony called Vinland.

But over the next few centuries, the Viking tide retreated. Perhaps their failure was due to climatic change or perhaps it was do to violent competition within Viking society. Certainly, the violent reaction of Native North Americans to Viking predations played a role in the demise of Norse America. Another possible contribution to their failure is Viking dependence upon pillage and trade, as opposed to the development of indigenous industries utilizing local resources.

A few centuries later, history witnessed on of the great "what might have beens" of all time. In the early fifteenth century, Ming Dynasty China constructed a huge fleet of enormous ocean-going junks and used these vessels to visit ports throughout Asia and Africa

With the support of Ming Emperor Yung-Lo, the Moslem eunuch admiral Zheng Ho used these junks—some of which measured 130 meters bow-to-stern and had crews of 500 men—to show the flag in Asian and African waters and ship novelties home to the court. Unfortunately, policy changed upon the death of the emperor; the fleet was recalled and the junks rotted at their piers. Although Imperial China certainly had the technology to spread its culture around the globe, the will for such an endeavor was lacking.

It was left to fifteenth-century European powers with much fewer resources than Imperial China at their disposal, notably Portugal and Spain, to initiate the Age of Discovery. Using the newly developed caravels, Portuguese navigators were encouraged by Prince Henry the Navigator to explore, fish, and trade farther from home. Some have speculated that the comparative scarcity of resources in tiny Portugal led to these explorations around Africa and to the far ports of exotic India.

Driven by greed and religious fervor (some would call it bigotry), Spanish Conquistadors followed Columbus to the New World in ships not derived from those of Prince Henry. Rather than settling in their new Caribbean and South American holdings, most Conquistadors sought to make it rich and return to Seville on the backs of native American and imported African slaves. It is a good thing that these slave societies were ultimately supplanted in the New World by the somewhat freer colonies of northern European powers.

After the American Revolution, the 13 former British colonies were huddled along the eastern coast of the North American continent. Thomas Jefferson proved to be one of the most visionary US presidents when he commissioned the Lewis and Clark expedition to begin the exploration of the vast continental interior.

The 13,000-kilometer trek of the Lewis and Clark "Corps of Discovery" began near St Lewis in May 1804, crossed the continent to the Pacific and ended in September, 1806. Since weight limitations were substantial, food supplies for the 48 men in the expedition were supplemented by hunting. With the assistance of friendly Native Americans, notably Sacagawea and her French-Canadian husband Toussaint Charbonneau, members of the Corps of Discovery were able to supplement their diets with local vegetation, thereby learning how to truly "live off the land.'

The success of the Lewis and Clark preliminary continental survey led to further exploration and the westward migration of the nineteenth century. Settlement would have been considerably slowed if efficient means of transporting people and baggage westward and frontier products eastward did not exist.

One efficient transport mode was the Conestoga Wagon. These "Prairie Schooners" had boat-shaped bodies topped with white sail-like canvas bonnets. Pulled by teams of horses, mules, or oxen, they could carry as much as 7,000 kilograms and were about 3 meters in length.

These wagons were equipped with tool kits so that repairs could be made *en route*. Such a provision was essential, since the nearest repair facility might be almost a thousand kilometers distant.

Although the Conestoga Wagon was instrumental in opening the American West, transcontinental transport using this method could not keep to a timetable. Human passengers found them very uncomfortable and they were hard on the animal teams providing the motive force.

A vast improvement on the Prairie Schooner was the Transcontinental Railroad, which was completed in 1869. Surveyed by the US Army Topographic Corps, this monumental project required the support of both the Federal Government and private interests.

As the examples presented above indicate, a number of factors are required for a successful human territorial expansion. These include innovation, the ability to live off the land, a flexible ideology, and a partnership between public and private sectors. Also necessary is at least a modicum of good luck and a good deal of determination!

FURTHER READING

An excellent, although somewhat dated treatment of early human evolution, authored by a Kenyan archeologist, is Richard E. Leakey's *Origins* (Dutton, New York, 1977), which was co-authored by Roger Lewin. Along with his father Louis and mother Mary, Richard Leakey contributed to our knowledge of hominid evolution in the Olduvai Gorge region of Kenya.

No wooden rafts used by the ancestors of the Aboriginal People to reach Australia have been preserved, but an excellent fictional portrayal of this migration is included in Stephen Baxter's science-fiction novel, *Evolution* (Ballantine, New York, 2003).

An excellent reference discussing the late prehistory and early history of humanity from 35,000 BC to the 500 AD is Jacquetta Hawkes' *The Atlas of Early Man* (St Martin's Press, New York, 1976). Of special interest is her description of how successful migrants adapted pottery styles, architecture, etc., to conform to their new environments.

Many references have examined the sea-faring contributions of the Minoan civilization in Neolithic and Bronze Age Crete. One very readable source is Rodney Castledon's *Minoans* (Routledge, New York, 1991).

Perhaps more than any other group of migrants, the Polynesians may serve as models for ultimate human expansion into the galaxy. You can read about their exploits in an article by Ben R. Finney, "Voyagers into Ocean Space," which has been published in *Interstellar Migration and the Human Experience*, edited by Ben R. Finney and Eric M. Jones (University of California Press, Berkeley, CA, 1985).

The same reference considers the expansion, and later contraction, of Viking culture in a chapter by Richard B. Lee entitled "Models of Human Colonization: !Kung, San, Greeks, and Vikings." The failure of Ming China and the success of fifteenth-century Portugal in initiating the Age of Exploration is also explored in this reference by Ben Finney, in a chapter entitled "The Prince and the Eunuch."

Many sources describe the Lewis and Clark expedition and the opening of the American West. Some of these are reviewed in a paper authored by Les Johnson, David Hardy, Ann Trausch, Gregory L. Matloff, Travis Taylor, and Kathleen Cutting, entitled "A Strategic Roadmap to Centauri." This article was published in *The Journal of the British Interplanetary Society*, vol. 58, pp. 316–325 (2005).

2

THE NEW FRONTIER

O'er the smooth enameld green,
Where no print of step hath been,
Follow me, as I sing
And touch the warbld string;
Under the shady roof
Of branching elm star-proof
Follow me.

John Milton, from *Arcades*

FAR above the emerald green seas that served as passageways through the Mediterranean, the Polynesian Pacific and to the New World, far above the tremulous canopies of the elm forest and beyond the finest wisps of Earth's life-giving atmosphere sits the Interplanetary Frontier. In crystal clarity, telescopes and space probes have returned photos of the strange environments of other solar-system worlds. Where in nearby space might human settlements best be established and how might humans and their descendents live in these strange new environments?

Broadly speaking, other than the Earth, there are three different types of environments within our solar system that will be of interest to early space pioneers. There are small, airless worlds such as our Moon, marginally habitable places such as Mars, and mountain-sized asteroids or comet nuclei. From the point of view of future human pioneers and settlers, each has advantages and disadvantages.

We omit from consideration such places as torrid Mercury and Venus and giant worlds such as Jupiter. Although far-future humans or their descendents might learn how to terraform Venus into a more clement, Earth-like world, or disassemble the gas giants to construct hordes of space cities, realization of such speculations lie very far in our future.

THE MOON AS A HABITAT

Because of its relative proximity, we begin our survey with Earth's single natural satellite, the Moon. Luna is the only solar-system world, other than the Earth, on which humans have walked. Many robotic probes have flown by, orbited, landed upon, or returned samples from the Moon. During 1968–1972, the crews of Apollos 8 and 10 orbited the Moon, Apollo 13 circled the Moon during its aborted mission, and the lunar modules of Apollos 11, 12, and 14–17 landed upon the lunar surface (Figure 2.1).

Following in the footsteps of the Apollo astronauts, future lunar pioneers will encounter an environment vastly different to any on Earth. The Moon orbits the Earth once every month, at an average distance of almost 400,000 kilometers. Contemporary rockets require about three days for a one-way lunar voyage.

Because the Earth has a mass 81 times that of the Moon, lunar gravity is much weaker than terrestrial gravity. Discounting the mass of the gear required to keep a person alive in the Moon's vacuum, a person weighs only one-sixth as much on the Moon's surface as on Earth.

Not only does the Moon *not* have an atmosphere, but water is either very rare or non-existent there. Two space probes—Clementine and Lunar Prospector—have returned preliminary data indicating that water-ice from ancient comet impacts may exist in craters near the lunar poles that are shielded from sunlight. But until this is confirmed and we have a better idea regarding the magnitude of this resource, future lunar colonists must plan to recycle as much of their habitat's water as possible.

FIGURE 2.1 Apollo 15 astronaut David Irwin works on the lunar rover during an excursion near Mt Hadley. Hollow volcanic lava tubes in this region might someday house human settlers. (Courtesy NASA)

Because the Moon does not have an atmosphere, one might think that solar energy would be more effective in powering a lunar civilization than it is on Earth. However, because of tides produced by the much more massive Earth, the Moon's rotation rate around its axis is identical to its revolution rate around the Earth. This means that a lunar "day" consists of 14 Earth days of sunlight followed by 14 Earth days of darkness. Because of this factor and the lack of an atmosphere, lunar temperatures are quite variable, ranging from a low of −170 degrees Celsius to a high of 130 degrees Celsius.

Unless we restrict human settlement to its polar regions which will benefit from near-continuous sunlight, any permanent lunar residents will have to contend with these temperature variations. Also, some form of energy storage facility or auxiliary power source will be necessary to supplement direct solar energy during the long lunar nights.

Some consider the lack of a lunar atmosphere to be a great advantage for future astronomical facilities based on our Moon. It has been suggested that a radio observatory on the lunar farside would be shielded from terrestrial radio signals by the Moon's limb. However, much larger radio telescopes could be constructed in free space and shielded from terrestrial radio signals by bulk material or simply pointed away from the Earth.

Another possible function for a future human civilization—mining of material for use on the Earth or in space—also has issues. Some have suggested that copious reserves of helium-3, a form of helium ejected from the Sun and of possible significance to a future nuclear-fusion-based economy, exist in usable concentrations in the lunar soil. Although this is possible, concentrations are very low and the Apollo core samples only extended a meter or so into the lunar dust and regolith. We may have to go further—the atmospheres of the giant planets, or the solar wind—to tap into significant helium-3 reserves.

Advocates of lunar mining and science may come into conflict. No matter how clean a mining process, quantities of fugitive lunar dust will be raised above the surface. In the low lunar gravity, these dust grains may remain aloft for many months, partially obscuring the pristine lunar skies sought by some astronomers and contaminating sensitive optical instruments as the dust inevitably falls back to the surface.

Even though volatile material such as water is lacking or very rare on the Moon's surface, methods have been suggested that could be used to catapult mined lunar rock from there. This material might be of use in constructing solar-powered satellites to beam energy to Earth or as cosmic-ray shielding for large space habitats. However, near-Earth objects (NEOs)—a class of asteroids and comet nuclei that approach the Earth— might be a source of higher grade ore and may require less energy to access economically.

LIVING AMONG THE NEAR-EARTH OBJECTS

Beyond the Moon, the next possible destinations for our space pioneers are the NEOs. The main source for these Earth-approaching objects, most of which do not exceed a kilometer in size, are the asteroids and comets.

Left over from the solar system's formation almost five billion years ago, asteroids are rocky and stony space objects typically found in the Asteroid Belt between the orbits of Mars and Jupiter. Gravitational perturbations have altered the positions of some asteroids—a number co-orbit with the giant planets or have been captured as their satellites, others have been thrown into Earth-crossing trajectories. The largest asteroid, Ceres, is 914 kilometers in diameter. Only about 150 of the 10,000 or so known asteroids are greater than 100 kilometers in diameter.

Earth-approaching asteroids tend to be much smaller. Astronomer John Remo estimates that there are about 20 NEOs with diameters greater than

5 kilometers, around 2,500 with diameters in the 1–2 kilometer range, and approximately 100,000 with diameters between 0.1 and 1 km.

A significant fraction of the NEO population is of cometary rather than asteroidal origin. Comets are multilayered objects left over from the solar system's origin. Around a rocky nucleus perhaps 20 kilometers in diameter are layers of volatile ices—water, methane, and ammonia. Many comets also have a layer of dust overlaying the volatiles. When a comet enters the inner solar system, solar heating evaporates some of the volatile ices. A typical comet making its inner solar-system passage is observed to have multiple tails of dust and volatiles, pushed by solar radiation pressure away from the Sun. Although a comet tail might be 100 million kilometers in length, it is so tenuous that all its material could be packed into a steamer trunk.

Comets usually reside in two reservoirs. Oort Cloud comets typically orbit the Sun at distances of a trillion kilometers or more. Those in the Kuiper Belt typically orbit a few billion kilometers from the Sun. The planet Pluto is the largest known Kuiper Belt Object (KBO).

Oort Cloud comets are scattered sunward during rare passages of other stars through or near the Oort Cloud. KBOs enter the inner solar system (sometimes taking up residence as NEOs) after gravitational perturbations by the giant planets, especially Jupiter and Saturn.

Since many NEOs approach the Earth at distances measured in millions of kilometers, much less energy is required to reach them than more distant celestial objects such as Mars. Trajectories have been designed that allow human exploration of some NEOs using current propulsion technologies with round-trip durations of a year or less.

Human colonists amid the NEOs will have to contend with the fact that these low-mass objects have surface gravities about one thousand times less than that of the Earth. To avoid bone degradation in the milligravity environment of a NEO, human colonists will be obliged to rotate their space habitat to produce artificial gravity. Radiation shielding for their habitat should be relatively easy to obtain from the upper NEO surface layers. Radiation is reduced or eliminated by placing lots of matter between the person and the radiation source. Since many NEOs have ample reserves of water and other volatiles, NEO colonists have less trouble with this requirement than their colleagues on the Moon.

In all likelihood, a human space habitat would use a NEO (or many NEOs) as a resource base. Unlike St Expury's *Little Prince*, NEO colonists will probably live near a NEO, or within a hollowed-out NEO, rather than on one.

Chapter 2

THE LURE OF MARS

Using chemical, solar, or nuclear rockets, 6–9 months are required for a one-way voyage to Mars. Fourth from the Sun and never closer to the Earth than 56 million kilometers, the Red Planet has baited human astronomers and fiction authors for centuries and tantalized space explorers for decades.

Early telescopic observers learned that Mars has polar caps, like the Earth. Some detected seasonal changes and occasional clouds. At the limits of resolution, a number of influential nineteenth-century astronomers reported a network of fine lines on the planet's surface—the legendary canals.

By the early twentieth century, these observations and science fiction of authors, including H.G. Wells and Edgar Rice Burroughs, firmly entrenched the idea of a Martian civilization in the human psyche. On this small and arid world, a fictitious ancient civilization had constructed canals to divert water from the polar regions to the equatorial cities. But as the atmosphere was escaping and the water reserves were emptying, Martian interplanetary imperialists began to turn their attention to the third planet from the Sun, our fertile Earth. Fortunately, it is now evident that the canals were illusory and that indigenous higher life forms do not exist on Mars.

Unlike the Moon with its month-long "day" and the NEOs with their widely variable rotation rate, the Martian "Sol" is only slightly longer (40 minutes) than the terrestrial day. But that's where the similarity between the two worlds ends.

Mars has an exceedingly thin atmosphere, with a surface pressure less than 1% that of the Earth and a surface gravity about 38% that of Earth. Although the polar caps do indeed have some water ice, their major constituent is dry ice—frozen carbon dioxide. This is not surprising since CO_2 is the predominant gas in the planet's thin atmosphere and frigid temperatures near the Martian poles are low enough to cause it to freeze.

Impact craters abound on Mars, as do enormous ancient volcanoes. One of these giants, Olympus Mons, dwarfs Mount Everest in both girth and height. Another major feature is Vallis Marineris, an equatorial canyon system far greater in size than North America's Grand Canyon.

Most tantalizing of all Martian surface features are the unique landforms produced by ancient rivers and seas. At some time in the distant past, Mars shimmered with abundant surface water, perhaps resembling a somewhat smaller version of the Earth.

Recent evidence from Martian rovers and orbiters indicates that some portions of Mars have been water-covered comparatively recently. Water may indeed be common in subsurface layers of the planet's regolith. One trace constituent of the Martian atmosphere is methane, which is a byproduct of terrestrial biological processes. This data, in combination with the ambiguous bacteria-like structures found in the mid-1990s in a meteorite from Mars, has led many exobiologists to the conclusion that Martian life was abundant in the past and may still exist in subsurface locations or caves today.

Future robotic and human expeditions to the Red Planet are planned, many of which will search for life on Mars or return samples directly to the Earth. Hopefully, we will know before too many decades have elapsed whether Gaia (Mother Earth) has a near neighbor in the solar system.

The search for Martian biology is a double-edged sword for would-be colonists of this planet. Early human expeditions to the Red Planet will almost certainly be driven by astrobiology. But if Martian life is actually discovered with a tenuous foothold in caves or below the planet's surface, a furious ethical debate is sure to ensue. Some will argue that Mars should be declared a preserve for indigenous Martians, no matter how primitive and that the human technological ability to cross the void is not a divine right to impose our biology on a distant biosphere. Others are sure to respond with the Darwinian argument that terrestrial life forms, having demonstrated the ability to survive interplanetary travel, are more fit to survive than the residue, stay-at-home Martians. Also, as happens on the Earth, where life that adapted to extreme environments such as the polar caps, deep-ocean floor, or upper atmosphere, can coexist successfully with our form of abundant surface life, perhaps Martian and terrestrial life forms can successfully share the Red Planet.

If people do ultimately settle on Mars, the first settlements may be shielded from cosmic radiation by Martian soil offering shelter not provided by the planet's atmosphere. The domed cities of science fiction may be constructed later as terraforming efforts begin the thickening of the atmosphere. After hundreds or thousands of years, perhaps with the aid of greenhouse gases imported from the Earth, the Martian atmosphere might be thick enough and the surface warm enough to support free-standing bodies of water, forests, parks, and outdoor agriculture. Although Mars may never be a twin of the Earth, it may ultimately become a second planetary biosphere in Sol's system.

Chapter 2

SETTLING THE OUTER SOLAR SYSTEM

Beyond Mars, there are ample opportunities for human expansion. The tiny Martian satellites, Deimos and Phobos, could be developed as refueling depots and way-stations for Mars-bound explorers and colonists. Similar to a class of asteroids resembling carbonaceous chrondite meteorites, these objects are probably extinct comet nuclei captured by Mars and, therefore, well equipped with water and other volatiles.

Farther out, between the orbits of Mars and Jupiter, are the main belt asteroids. These will serve as resource bases for human space settlements as human civilization expands into the outer solar system.

Next out from the Sun is the giant planet Jupiter, which has more than 300 times the mass of the Earth. Jupiter, a very strong radio emitter (strong enough, in fact, to be heard on Earth in the shortwave radio band at 20.1 megahertz), is encircled with radiation belts more lethal than the Earth's Van Allen Belts. It's four large satellites—Callisto, Europa, Ganymede, and Io—are interesting and exciting worlds with unique surface features. Io has giant, continuously erupting volcanoes that alter the landscape at a prodigious rate. Europa is apparently encircled by a water ocean many kilometers deep and covered with an ice pack—tantalizing space probe imagery reveals that some parts of this ocean are not frozen. Ganymede and Callisto have features indicating both water deposits and impacts by smaller celestial objects. Sadly, Jupiter's intense radiation belts may preclude widespread human settlement among these interesting little worlds.

The sixth world from the Sun, Saturn, is like all the giant worlds, attended by a system of rings that were probably formed when asteroids or comets approached the planet too closely. But unlike the rings of Jupiter, Uranus, and Neptune—which are only visible from space probes or space telescopes—Saturn's rings are easily viewed using binoculars or small telescopes. Although we may never attempt to live in the alien environments of Saturn space, the Huygens probe recently touched down on the surface of Saturn's large satellite Titan, revealing a bizarre landscape perhaps not dissimilar to the primeval Earth.

Next out are the smaller giant worlds of Uranus and Neptune. It remains to be seen whether we will ever live on the satellites attending these majestic gas balls in the frigid fringes of the solar system, but some day we might mine their atmospheres for isotopes such as helium-3, which will be of great value to a thermonuclear-fusion-based economy.

Beyond Neptune we come to the realm of the Kuiper Belt Objects (KBOs). Pluto, which requires almost 250 years to complete one revolution around the Sun, is one of the largest known KBOs.

The KBOs contain ample reserves of frozen water, methane, and ammonia. Therefore, intrepid far-future humans equipped with low-mass solar collectors to focus the emanations from the distant Sun could, with effort, settle these small, frigid worldlets. Most of the required technologies for such expansion will have been tried among the much closer NEOs and main belt asteroids. Additional interest in exploration is generated by evidence that one source for the short-period comets that occasionally whack the Earth with devastating consequences is the Kuiper Belt.

Beyond the Kuiper Belt cometoids is the Oort Cloud, the home range of perhaps a trillion comets. The Oort Cloud comets, some of which range a third of the distance to the nearest stars, are the reefs of Sol's system, lonely sentinels on the edge of infinity. When humans are technically ready to settle this distant realm, their space habitats must be essentially independent of distant Terra. At that juncture, we will also be capable of traversing the greater distances to the nearer stars.

INTERSTELLAR ENVIRONMENTS

Recent telescopic evidence suggests that many or most stars have planetary companions. Although we cannot reliably detect Earth-like worlds, a multitude of extrasolar Jupiter-like worlds have been discovered. Many of these new worlds have bizarre orbits—circling so close to the parent star that the orbital period is but a few days, or riding extreme comet-like ellipses. But some of the newly discovered solar systems superficially at least, resemble our own. Since both single and binary stars have been observed to possess planetary companions, many of our Sun's stellar neighbors are potential habitats for future star-traveling humans.

The Sun's distance from the Earth is 150 million kilometers or 1 Astronomical Unit (1 AU). Pluto's average distance from the Sun is about 40 AU. If we some day learn to build ships capable of crossing a 260,000 AU (or 4.3 light-years) interstellar gulf, our descendents will be capable of reaching our Sun's nearest stellar neighbor, the triple star system called Alpha Centauri.

Interstellar explorers may choose to ignore the closest member of this trio, Proxima Centauri. A red dwarf star with very low luminosity, Proxima is probably too dim to support life-bearing worlds.

Such is not the case for the two central sunlike stars, Alpha Centauri A and B. The separation between these stars is ample for one or both to have a solar system. Our nearest stellar neighbor may in fact have more than one Earth, and more than one Mars. Many satellites, asteroids, and comets may also grace the Centauri system.

Ships that can traverse distances of 11–12 light-years could explore the systems of Epsilon Eridani and Tau Ceti, our next nearest solar-type stellar neighbors. Although recent observations indicate that Epsilon Eridani may be attended by a Jupiter-sized planet, this system may be too young for life to have evolved on any of its planets or satellites.

As recent observations have revealed that Tau Ceti is surrounded by a very extensive asteroid belt, repeated asteroid impacts may preclude the evolution of advanced life in this system.

We are still new to the game of extrasolar-planet detection. As larger or more capable space and terrestrial telescopes come on line, we will certainly learn a good deal about the environments that await us among the stars.

We have come a long way in the 2,500 years or so of astronomical observations. At the start of the Athenian golden age, the planets were divine entities and the stars were portals to heaven. Now, the planets are worlds more or less like our own and the stars are Suns with their own solar systems. Life may well exist in the reaches of this New Frontier or may be carried there by ships from Earth.

FURTHER READING

Prospects for further lunar exploration and lunar habitation are discussed by E. Burgess, in *Outpost on Apollo's Moon* (Columbia University Press, New York, 1993). An excellent introduction to the Red Planet is M. Caiden and J. Barbree with S. Wright, *Destination Mars* (Penguin Putnam, New York, 1997).

Many college-level astronomy texts summarize current knowledge of small solar-system bodies and outer-solar-system environments. Especially notable and readable is E. Chaisson and S. McMillan, *Astronomy Today*, 3rd edn (Prentice-Hall, Upper Saddle River, NJ, 1999). For the most up-to-date information on solar-system exploration, consult popular astronomy magazines such as *Astronomy* and *Sky & Telescope*.

Prospects for free-space habitats is discussed by G.K. O'Neill in *The High Frontier* (Morrow, New York, 1977). Use of solar-system resources to support these future facilities is further investigated by B. O'Leary, in *The Fertile Stars* (Everest House, New York, 1981).

Lots of information on planets of other stars is included in P. Gilster's non-technical *Centauri Dreams* (Copernicus, New York, 2004), which also discusses other possible means of reaching the stars.

$$MR = Exp(delta\,V/Vex)$$

Surface-Running Torpedo

3

THE ROCKET AND ITS LIMITS

When the Powers of the Air are chained to my chair,
*   Is the million-colored bow;*
The sphere-fire above its soft colors wove,
*   While the moist Earth was laughing below.*

Percy Bysshe Shelley, from *The Cloud*

IT should not be surprising that chemical rockets derive their thrust from chemical reactions. The reaction between a rocket fuel and an oxidizer (a chemical compound containing oxygen) releases a tremendous amount of energy. The energy is released as new chemical bonds are formed in the "burning" process. Channeling the energetic reaction products outward from the vehicle using a directional nozzle produces thrust. The hot gas goes one way, the rocket ship the other, and momentum is conserved. Chemical rockets are propelled by reactions of either liquid or solid fuel combinations. Solid fuels are generally easier to store for long periods of time; liquid propellants are typically more energetic.

29

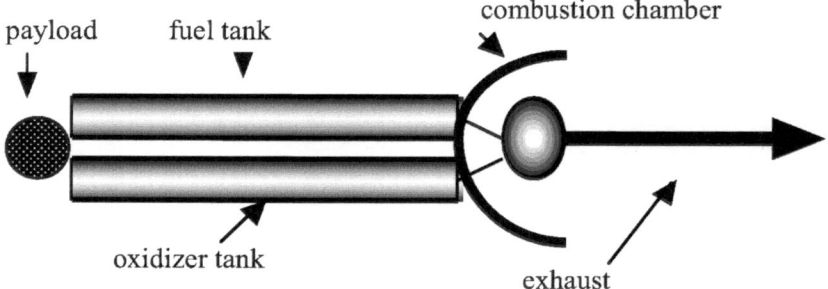

FIGURE 3.1 A chemical bipropellant rocket.

The chemical reactions that drive a rocket are called "exothermic," meaning that these reactions release more chemical energy than they require. Most chemical rockets (as shown in Figure 3.1) utilize bipropellants—a fuel and an oxidizer. In the liquid engines of the space shuttle, the fuel is liquid hydrogen and the oxidizing agent is liquid oxygen.

Some chemical fuel combinations, such as hydrogen and oxygen, react explosively on contact. Others must be preheated to react.

Rocket designers spend a great deal of time optimizing such elements as valves, nozzles, and pumps. But barring breakthroughs in the field of exotic chemical fuels, the space shuttle main engines' 4.55-kilometer-per-second exhaust velocity is close to the theoretical maximum. It cannot get much better than that.

Although rockets were first applied to space travel during the 1940s and 1950s, the history of the Rocket Principle is a good deal more venerable. Hero of Alexandria may have been the first rocketeer in the first century AD when his "aeropile" used jets of steam to rotate a suspended structure.

However, almost a millennium would pass until the first rocket-propelled devices, the "fire arrows" of the Chinese inventor Feng Jishen, would streak through Earth's lower atmosphere. These hollow bamboo tubes had holes at one end and were filled with gun powder. With a bamboo stick attached for stability, the early fire arrows could climb into the sky, producing the ancestor of modern firework displays.

According to legend, the first attempt at manned rocket flight may long predate the twentieth century. Before the year 1500, a Chinese mandarin named Wan Hu, perhaps world weary or a bit mad, directed his retainers to attach a large number of fire arrows to a chair, which was also equipped with a few kites to, perhaps, aid his atmospheric cruise. Wan Hu sat in the chair, which had been dragged outside. His retainers lit the fuse and probably ran as fast as possible. This intrepid (or foolhardy) pioneer disappeared in the ensuing explosion. Perhaps Wan Hu lived long enough to view the moist, blue Earth below his spinning rocket chair; but he more likely died instantly.

Early rockets were not capable of precision flight but they were certainly impressive. Their earliest application in Asian warfare probably was as terror weapons to demoralize opposing armies with loud (and unpredictable) explosions.

When Marco Polo reinvigorated east–west trading contacts at the dawn of the Renaissance, one concept that moved west was the Rocket Principle. Although used in western firework displays, the rocket does not surface as an effective European weapon until the eighteenth century.

This delay was most likely due to the difficulty in controlling the rocket's post-launch trajectory. Perhaps the most successful military rocketeer of that era was Sir William Congreve, a British artillery officer. His specially constructed "rocket ships" were used as an effective siege weapon during the Napoleanic wars. One has been immortalized in American patriotic music by the phrase "by the rockets red glare."

A number of non-military rocket applications were also experimented with during the eighteenth century. One was an effort by Prussian and British rocketeers to develop a line-carrying rocket that, after launch from a lighthouse, could become entangled in the sails of a foundering ship, allowing crew and passengers to scamble to land.

With all the nineteenth-century experimentation, it is surprising that early science-fiction writers chose huge guns rather than rockets to propel their spacecraft through interplanetary space. But the first person to seriously consider the rocket for space propulsion was a visionary scientist rather than a fiction author.

Konstantin Tsiolkovsky (1857–1935) was born in Czarist Russia and lived to see his theoretical rocket researches widely acknowledged. In his spare time, this mathematics teacher composed treatises that dealt with weightlessness, the problems of maintaining closed ecological systems in space, the capabilities of various chemical rocket fuels, and the concept of staged rockets. He successfully identified liquid oxygen and hydrogen, the fuel of the American space shuttle's liquid rockets, as an extremely energetic propellant for chemical rockets. Tsiolkovsky was perhaps the first to realize that high exhaust velocity was the key to rocket-propelled interplanetary travel.

Although Tsiolkovsky was a theoretician who never actually constructed a rocket, his work inspired further research. One of his successors, the Romanian Hermann Oberth (1894–1992), published best-selling books on the potential of the rocket that led to the formation of German amateur rocket societies. The experiments of these groups, principally the "Verein für Raumschiffarht," were later incorporated into Nazi-sponsored research leading to the development of the V2. Although principally applied as a terror weapon capable of bombarding London from launching pads hundreds of kilometers away, the V2 was the first terrestrial vehicle that rose above the Earth's atmosphere.

Robert Goddard (1882–1945)—an American contemporary of Tsiolkovsky and Oberth—was a physics professor with a strong interest in experimentation. He constructed a number of liquid-fueled rockets and launched them from sites in Massachusetts and New Mexico. He was granted more than 200 patents for his rocket research and described much of it in his 1919 publication *A Method of Reaching Extreme Altitudes.* Goddard's innovations included the application of gyroscopes to rocket guidance, use of vanes in the exhaust stream for steering, and valves to control propellant flow.

Oberth and Tsiolkovsky were lionized by the Russian and German public and political authorities. Goddard, on the other hand, was ridiculed by the American press and his experimental launches were banned in Massachusetts.

After the Second World War, the victorious powers realized that an ultimate weapon might be a ballistic missile carrying a nuclear device as its payload. Thus began the "space race," as the USA and the USSR sought to improve missile reliability, range, and payload capacity and to develop miniaturized electronics for guidance computers and nuclear payloads. Most American and Russian space launches prior to the Apollo Moon-landing program utilized converted or uprated war rockets.

Utilizing these craft as they competed for world technological dominance, the space powers chalked up an impressive list of space accomplishments in the first decade of the Space Age. Early satellites orbited the Earth at a variety of altitudes, where they discovered such aspects of the terrestrial environment as Earth's Van Allen radiation belts. During the first few years of the Space Age, crude robots impacted the Moon, followed by others that orbited and touched down on the lunar surface. The exploration of the solar system moved into high gear as probes journeyed to Earth's neighboring worlds Venus and Mars.

Although it is largely taken for granted today, the commercial space infrastructure is another accomplishment of pioneering space missions. Before 1970, weather, communication, and Earth-viewing satellites had made their appearance.

But the public's attention, of course, was concentrated upon the human drama of the Moon Race between the USA and the USSR. Those old enough to witness it will never forget how they held their breath as early astronauts and cosmonauts flew ballistic trajectories to the fringe of space, orbited the Earth, and began to reach for the Moon. Along the way, they learned how to rendezvous and dock in space using rockets, to ship components up from Earth to be used in the construction of large space stations, and ultimately to develop a permanent human presence above the Earth's atmosphere.

ROCKET FUNDAMENTALS

It doesn't take a "rocket scientist" to understand rocket propulsion. It's all about conservation of momentum. Modern rockets come in three basic varieties: chemical, electric, and nuclear. In all three cases, an energy source excites a reaction mass, or fuel. This is expelled from the rear of the rocket at high velocities. The reaction to this fluid release manifests as thrust and pushes the rocket forward. The momentum of the *system* remains unchanged. If you add the momentum of the now–moving rocket to that of its exhaust, the sum will be zero. Rockets don't create something from nothing!

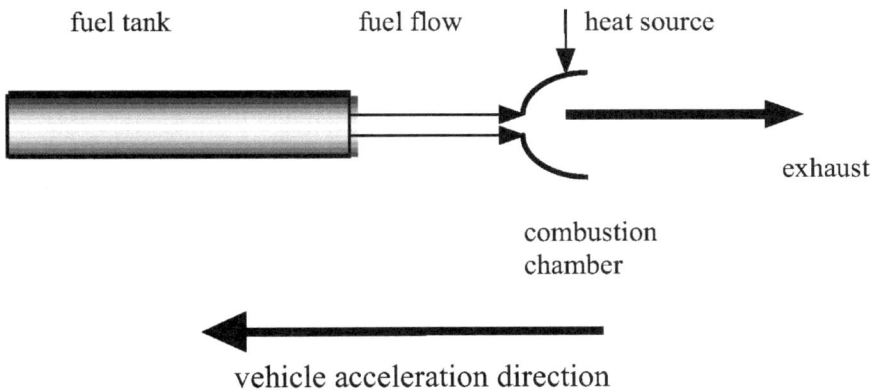

FIGURE 3.2 The rocket principle.

Figure 3.2 presents a generic rocket. The total mass of the fuel is M_f and M_e is the mass of the vehicle (engine, structure, fuel tanks and payload), exclusive of fuel. The ratio of fueled mass to unfueled vehicle mass [$(M_f + M_e)/M_e$] is called the Mass Ratio (MR).

If the fuel is expelled at an exhaust velocity V_e relative to the vehicle and the total change in velocity after all fuel is expelled is ΔV, application of Newtonian mechanics allows one to derive the famous Rocket Equation:

$$MR = e^{\Delta V/V_{ex}}$$

where e is the base of the natural logarithm system, which has an approximate numerical value of 2.72.

For the non-mathematician, the significant factor about this equation is the fact that it is non-linear, meaning that the Mass Ratio (and hence fuel) increases very rapidly as the velocity ratio $\Delta V/V_{ex}$ increases. If this velocity ratio is exactly equal to 1, the Mass Ratio is 2.72. If it is equal to 2, the Mass Ratio becomes 7.39 and if the velocity ratio is raised to 3, the Mass Ratio becomes 20.1 To accelerate to high velocities, rocket fuels must be expelled at the highest feasible exhaust velocities and the Mass Ratio must be as high as economics and technology permit.

To gain an idea of rocket potentials and limitations, the Rocket Equation can be used to determine the Mass Ratio for the case of the best performing chemical rocket used to put a spacecraft in orbit. We assume that the exhaust velocity is 4.55 kilometers per second, about equal to that of the space shuttle's liquid fuel rocket engines in a vacuum and close to the theoretical maximum for a chemical rocket. To attain low Earth orbit (LEO), the vehicle must be accelerated to a velocity of 7.91 kilometers per second. (For readers not attuned to the metric system, this velocity is equal to 4.9 miles per second or about 17,700 miles per hour.)

The Earth rotates on its axis at a velocity of about 0.46 kilometer per second, so if the spacecraft is launched in an easterly direction from a near-equatorial site, it can gain a small velocity advantage from Earth's rotation. But its engines must still provide an incremental velocity of $7.91 - 0.46 = 7.45$ kilometers per second to achieve LEO.

If it is required to place the entire mass of the vehicle at lift-off into orbit, the Rocket Equation can be used to calculate the Mass Ratio. For this case, the Mass Ratio is 5.14. More than 80% of the spacecraft's mass on the launch pad is fuel.

Let's next consider the case of direct launch to Earth escape, say on a Moon-bound or Mars-bound trajectory. The total velocity increment for this case is 11.19 kilometers per second. With an easterly launch from an equatorial site, the rocket engines must accelerate the spacecraft to 10.73 kilometers per second (or about 24,000 miles per hour). Assuming the same exhaust velocity as before, the Rocket Equation can be used to calculate that the required Mass Ratio is 10.57. In this case, about 91% of the initial spacecraft mass on the launch pad is fuel.

The non-linearity of rocket operations is evident when one compares velocity and fuel requirements for Earth-orbital and Earth-escape missions. The velocity increment required from the rocket for Earth escape is 1.44 times the velocity increment required to achieve LEO. The Mass Ratio for Earth escape is, however, 2.06 times the Mass Ratio required to achieve LEO.

At first glance, these numbers do not appear too daunting, at least for LEO operations. After all, a rocket composed of 80% fuel does not seem too ridiculous.

In practice, however, very large mass ratios are not possible for Earth-launched spacecraft. A rocket cannot consist of only fuel and payload, for instance, because of the requirement for supporting structure as it accelerates at multiples of Earth's surface gravity. A very large fraction of a rocket's fuel is exhausted, in fact, in the first few seconds after it clears the launching pad. This is why the designers of the Pegasus booster, the X–15 and Space Ship 1 have opted for air launches. Not only is a rocket dropped from a high-altitude jet moving at a high transonic velocity; it starts its flight at an altitude high enough for a significant reduction in atmospheric drag.

A more commonly improved technique to improve rocket performance is staging. As a spacecraft climbs toward the fringes of the atmosphere, it usually drops exhausted engines and fuel tanks rather than continue to carry the unneeded dead weight.

Even though staging has allowed humans or robotic probes to reach orbit, the surface of the Moon, the vicinity of other solar-system worlds, and the fringes of interstellar space, it is a horribly uneconomical process. Typical launch costs are in the vicinity of thousands of dollars per kilogram. The process has been compared to what air travel would be like if a jumbo jet discarded various components during each flight which then had to be replaced after each landing.

One attempt to apply economics has been the development of partially reusable spacecraft such as the US space shuttle. But the shuttle must be significantly refurbished after each round trip into space, it requires an army of technicians to maintain it, and it has not proven to be an accident-proof method of entering the space frontier.

The shuttle also has a very small payload mass fraction. Sitting on the launch pad, the fully assembled shuttle masses approximately 2 million kilograms (4.4 million pounds). The maximum payload capacity is around 30,000 kilograms (65,000 pounds). The empty mass of the orbiter (not including payload and fuel) is 75,000 kilograms (165,000 pounds)—more than 90% of the shuttle's mass on the launch pad is either dropped in the ocean or discarded as fuel.

ROCKET VARIETIES

A number of rocket varieties either are in the operational inventory or have been designed. By far the most common is the chemical rocket. Solar-electric rockets have flown on interplanetary voyages. Solar-thermal, nuclear-electric, and nuclear-thermal rockets have been ground tested.

Electric Rockets

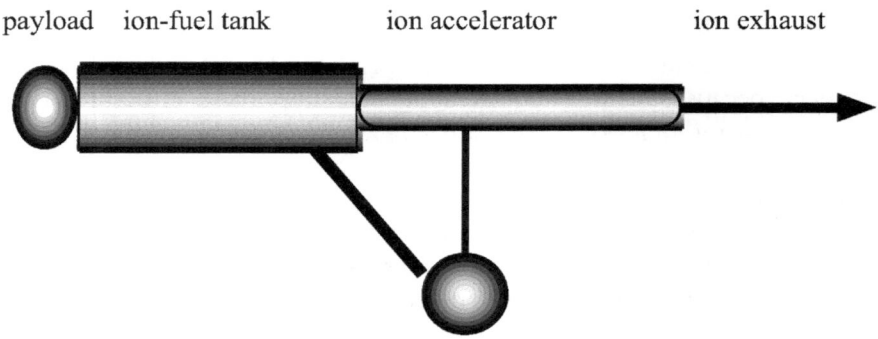

FIGURE 3.3 An electric rocket.

As shown in Figure 3.3, the electric rocket (also called the "ion drive") uses an external source of energy—solar arrays or an onboard nuclear reactor—to ionize reaction fuel and then to electrically accelerate the ionized fuel to a high velocity. As the high-speed ionized fuel leaves the spacecraft in one direction, the spacecraft moves in the other direction and, again, the momentum of the system remains unchanged.

Solar-electric engines are now operational, after having passed their space-qualification tests. Although capable of only very small accelerations (in the vicinity of 10^{-4} Earth gravities), ion-drive exhaust velocities can exceed 30 kilometers per second.

Due primarily to their low thrust, no ion-drive spacecraft will ever lift off the Earth into space, but given enough fuel and acceleration time (there is plenty of time when the destination is an asteroid many millions of kilometers away) they have demonstrated the capability of accelerating spacecraft to very high interplanetary velocities.

Solar-electric engines will find significant application in the inner solar system where sunlight is most intense, unless methods of beaming energy to interplanetary spacecraft are developed. For deep space missions far from the Sun, onboard nuclear power will almost certainly be required to power any ion engines. Today's technology falls short of that required to safely build, launch, and operate a nuclear-powered electric propulsion vehicle affordably in deep space.

Nuclear-Thermal and Solar-Thermal Rockets

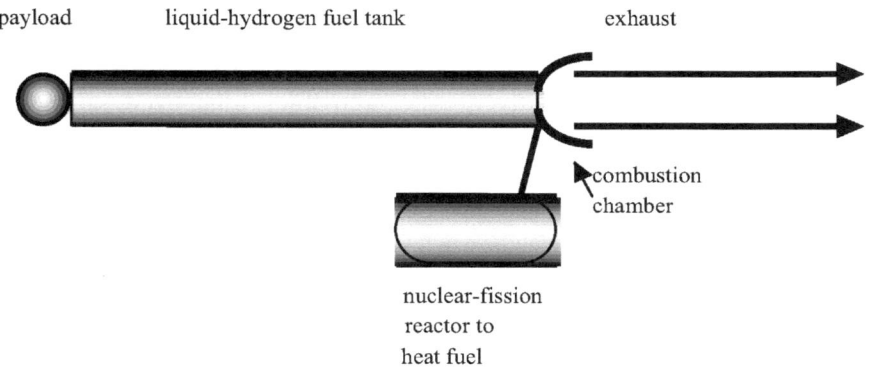

FIGURE 3.4 A nuclear-thermal rocket.

Nuclear-thermal rockets operate as shown in Figure 3.4. Here, the energy released from an onboard nuclear reactor is used to heat the fuel (usually liquid hydrogen) and expel it at high exhaust velocities.

A number of nuclear-thermal rockets were successfully ground tested during the 1960s and 1970s. If the nuclear reactor power levels are in the 100–1,000 megawatt range, these engines are capable of high-thrust operation at exhaust velocities about twice that of the space shuttle's liquid fuel chemical rockets. Nuclear-thermal rockets have the potential to be used for Earth-to-orbit applications as well as in space. However, many technical and safety issues must first be addressed.

The solar-thermal rocket replaces the onboard nuclear reactor with solar concentrators to superheat the propellant in much the same way as using a magnifying glass and sunlight can be used to burn a piece of paper or start a fire. Many of the technologies required for a nuclear-thermal rocket are applicable to a solar-themal rocket and will likely be first demonstrated using this non-nuclear propulsion system.

Ground tests have revealed that exhaust velocities as high as 10 kilometers per second are possible for the solar-thermal rocket. However, since the Sun is a diffuse energy source, solar-thermal rockets are low-acceleration devices best suited for in-space application.

ROCKET FUTURES

Though they have been the workhorses of the Space Age (so far), chemical rockets are a dead end for the species if we truly desire to explore and then settle the solar system. Although chemical rockets will certainly never go completely out of style, some other techniques are required for the large-scale opening of the space frontier.

Chemical rockets are not readily reusable and are quite dangerous (a chemical rocket is really nothing more than a bomb that "goes off" slowly and in a usually controlled fashion). Clearly, alternatives are needed. Ideally, those that are relatively safe, reusable, long-lived, and take advantage of the resources provided by nature throughout the solar system will be the next that humanity chooses to use as it expands beyond the Earth.

FURTHER READING

The legend of Wan Hu's early attempt at rocket flight is included in the classic book by Carsbie C. Adams, *Space Flight*, which was published by McGraw-Hill in 1958. A more recent monograph devoted to rocket history and theory, *Rocket and Spacecraft Propulsion* (2nd edn), was authored by Martin J.L. Turner and published by Springer–Praxis, Chichester, UK, in 2005. This reference is also an excellent source of information regarding chemical, electric, and nuclear rocketry.

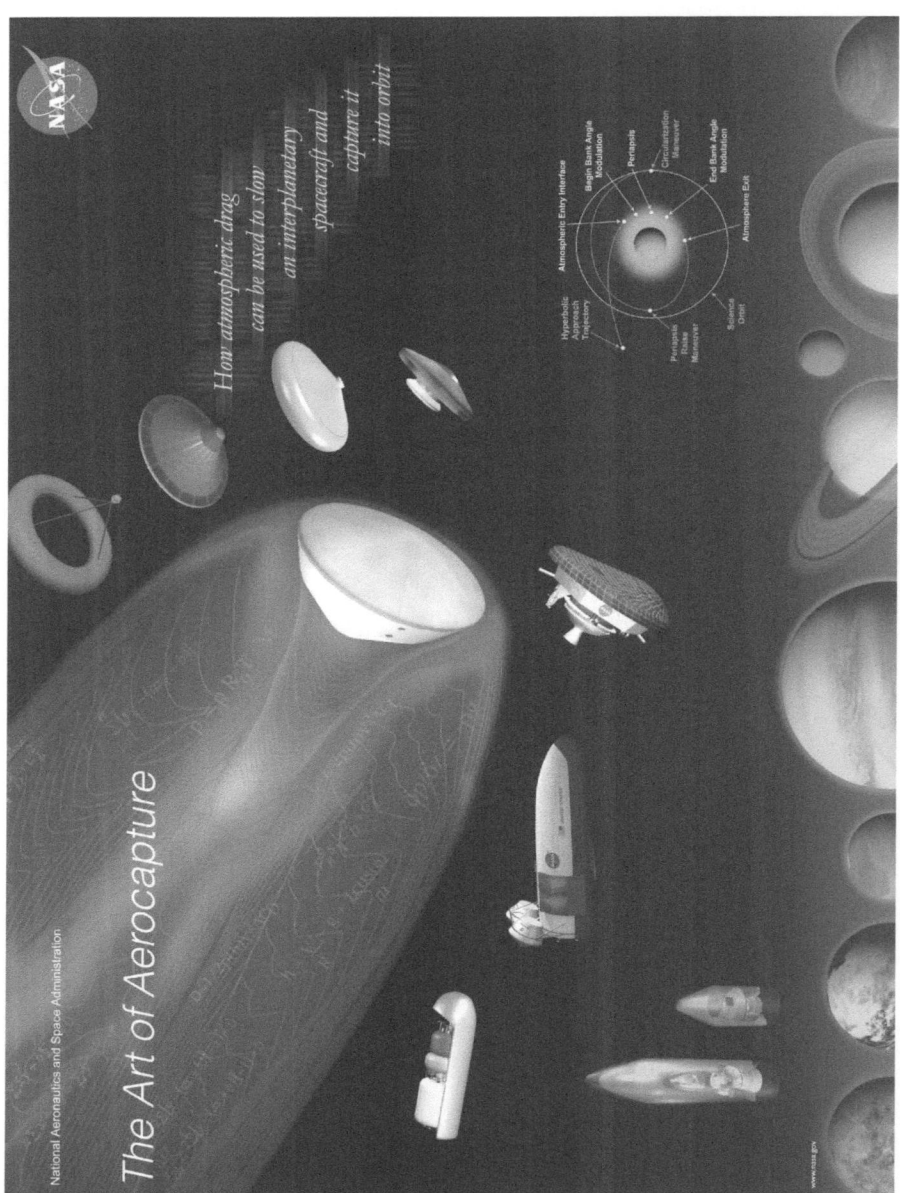

The Art of Aerocapture

How atmospheric drag can be used to slow an interplanetary spacecraft and capture it into orbit

National Aeronautics and Space Administration

NASA

[See also Plate 1 in the color section]

4

THE FIRST "GREEN" SPACE TECHNOLOGIES

And silken-sharp the dazzling thunder falls
Upon the startled land. The rising, falling dart
Sudden and piercing on the summer's heart;
And while from tree to tree the voice of fore calls

Marya Zaturenska, from *The Tempest*

EARLY in the Space Age, as thunder fell earthward from the rising rockets as they pierced the firmament, researchers realized the necessity of developing space technologies to exploit the Near-Earth environment. To get into space astronauts and cosmonauts would use the high-technology rocket. But to return from orbital journeys, spacefarers would use frictional drag produced by Earth's envelope of air. Another "green" space technology would also soon be developed—to tap the gravitational fields of giant planets.

Aeroassisted Atmospheric Reentry

One might think that a viable means of returning an Earth-orbiting spacecraft to our planet's surface could be effectively done by simply firing the rockets in reverse. But if you consult the Rocket Equation in Chapter 3, you will see that such an approach simply could not work.

In the example given in Chapter 3, the exhaust velocity of a near-optimum chemical rocket is 4.55 kilometers per second, and the velocity increment required to achieve minimal Earth orbit is about 7.45 kilometers per second. Substitution in the Rocket Equation reveals that a mass ratio of 5.14 is required to achieve a minimal Earth orbit. About 80% of the spacecraft's liftoff mass is fuel and oxidizer.

Now let's assume that enough fuel is required so that the rocket can exactly cancel its orbital velocity relative to Earth's surface. The velocity increment is now 14.90 kilometers per second. Solution of the Rocket Equation reveals that the minimum required mass ratio is now about 26.4. In this case, more than 95% of the rocket's lift-off mass consists of oxidizer and fuel.

If such large mass ratios were the only way to achieve earth-return, it is very doubtful if large, people-carrying spacecraft would ever have orbited the Earth, let alone ventured to the Moon. But our planet's atmosphere provides another solution.

Astronomers had actually been aware of the velocity-braking properties of Earth's atmosphere for centuries. Every time Earth intersects a comet's solar orbit, it encounters tiny grains of dust and ice from the comet's tail. As these grains approach our planet, they decelerate as they collide with air molecules in Earth's upper atmosphere. During these encounters, the comet grains' energy of motion is converted to heat. The grains begin to glow visibly as they streak across the sky and most of them evaporate at altitudes of 50 kilometers or so above Earth's surface.

The first terrestrial to enter orbit was the dog Laika, on board Sputnik 2 in November 1957. Because the problems of reentry had not been addressed at this early date, Laika became the first space casualty. She died in orbit after about one week in space.

The safe return of astronauts and cosmonauts from orbital missions was not the only driver to reentry research. Early in the Space Age, both American and Soviet mission planners realized the utility of returning high-resolution film to Earth in capsules released from spy satellites. Robotic reentry has also been used in a wide variety of scientific missions.

Both space powers arrived at the same general approach for returning astronauts or cosmonauts from orbit. At the conclusion of the flight, a space capsule is oriented to enable a small retrorocket to be fired in a direction opposite to the spacecraft's velocity. The capsule then descends until it encounters the outer fringes of Earth's atmosphere. As the craft descends into denser air, more and more of its energy of motion is converted into heat through friction between it and the molecules of air.

Spacecraft reentry is demanding in terms of trajectory. If a vehicle enters the atmosphere too steeply, it may burn up. If the reentry angle is too shallow the spacecraft may "surf" along atmospheric layers and skip back into space.

The aft end of the space capsule is coated with a special ablative shield. As friction with atmospheric molecules raises the temperature of the reentry shield to thousands of degrees Celsius, bits of the shield flake off or ablate. Much of the heat generated during reentry is carried away by the evaporated shield material, much to the relief of the crew!

At an altitude of perhaps 30 kilometers, the ship's velocity is low enough and the atmospheric is thick enough to enable aerodynamic surfaces and parachutes to be used for further deceleration.

Some contemporary spacecraft, notably the American space shuttle, are coated with temperature-resistant tiles rather than a solid shield, but as many of the shuttle's tiles must be replaced after each mission, this is a time-consuming process that contributes to the high cost of space-shuttle missions.

Although some Russian and American crews have died during reentry, shields or tiles have an otherwise perfect safety record during peopled space flights. The April 1967 the reentry death of Vladimir Komarov on board Soyuz 1 was due to failures in the attitude control system that caused the spacecraft to spin so violently that the parachutes could not deploy. A few years later, the crew of Soyuz 11 performed a flawless reentry—but the crew was found to be dead in the capsule because of a faulty air seal. The more recent loss of space shuttle Columbia was due to structural damage occurring when ice and foam flaked off the external tank during the ascent to orbit.

Astronauts and cosmonauts do not easily relax during their wild return to Earth. It's a bumpy ride and the realization that only a thin ablative shield or tile layer separates them from ionized gases at temperatures of thousands of degrees cannot be good for digestion. The meteor-like trajectories of returning spacecraft are impressive displays, attracting the attention of hordes of amateur astronomers.

PLANETARY GRAVITY ASSISTS

A different form of "green" space technology saw its first utilization during the 1970s when NASA launched the first probes toward the outer planets and stars. These were Pioneers 10 and 11 and Voyagers 1 and 2. These craft flew past all of the giant planets of the solar system—Jupiter, Saturn, Uranus, and Neptune—and have continued to radio back data from the edge of interstellar space. And they could not have succeeded without application of a space-propulsion technique that takes advantage of the environments of these giant worlds—planetary gravity assists.

As mission planners began to consider sending probes to the outer solar system, it was soon realized that conventional thinking simply would not do. Let's say we wish to launch a rocket to Jupiter along an energy-efficient trajectory. One approach is to fly along a so-called Hohmann transfer. In such a trajectory, the part of the elliptical transfer orbit closest to the Sun (the perihelion) is tangent to Earth's solar orbit. The part of the transfer orbit closest to the destination planet (the aphelion) is tangent to the solar orbit of the destination world. A spacecraft traveling along a Hohmann transfer to Jupiter must depart the Earth with an excess velocity of 8.57 kilometers per second relative to the Earth (this is called the "hyperbolic excess velocity"). The duration of the voyage is about 2.74 years. If we wish to use the same approach to visit Saturn, the spacecraft must depart Earth with a hyperbolic excess velocity of about 10 kilometers per second and the flight duration is about 6 years. To perform a Hohmann-transfer flyby of Neptune, the mission planners must be patient indeed—the voyage duration would be approximately 46 years.

Fortunately for humanity's deep-space aspirations, there is another approach. If a spacecraft flies by a giant planet along just the right path, it can "borrow" some of the giant world's orbital energy. The spacecraft departs its encounter with the giant world at a higher velocity relative to the Sun. Thanks to the mass difference between the two objects (the small spacecraft and the massive planet), the planet's heliocentric velocity is reduced by an inconsequential amount.

This sounds like magic, but it is actually a consequence of a basic principle of Newtonian physics, the Conservation of Angular Momentum.

Everybody who has played a game of billiards or pool has experienced a related physical principle, the Conservation of Linear Momentum.

The linear momentum of an object is defined as the product of the object's mass (M) and its velocity (V). If you consider an isolated system, say a cue ball (c) and a billiard ball (b), the total linear momentum of the system is $M_cV_c + M_bV_b$. In any interaction between the two balls, you can rearrange their individual linear momentum. But the total linear momentum of the two-ball system remains constant.

Let's say that you are taking careful aim on a billiard ball to strike it head-on with the cue ball in order to sink the billiard ball in a pocket. You first impart linear momentum to the cue pool with the pool stick. Upon impact, the linear momentum of the cue ball is transferred to the billiard ball. The cue ball becomes stationary after the collision and the billiard ball moves off until it enters the pocket.

The angular momentum of a planet of mass M in a circular orbit a distance R from the center of the Sun is expressed as MVR, where V is the planet's solar-orbital velocity. The total angular momentum of a system is also conserved. If you watch a figure-skater or ballerina in a fast spin, she can slow down by stretching out her arms. Since the distance R of her hands from the center of curvature increases in this manner, her velocity slows.

In a three-body system composed of a spacecraft and planet both orbiting the Sun, the individual angular momentum of the spacecraft and the planet can be rearranged, but the total angular momentum of the system is constant. During an unpowered gravity-assist maneuver, the spacecraft departs the planet's gravitational influence at the same speed relative to the planet that it had during its approach to that world. The spacecraft's trajectory direction, however, is altered by the gravity-assist maneuver, as is its velocity relative to the Sun.

A gravity-assisted mission to an outer-solar-system or extrasolar destination might proceed as follows. The spacecraft is inserted into a Jupiter-bound trajectory after leaving the Earth. If it flies by the giant planet in just the right manner (as illustrated in Figure 4.1), some of Jupiter's angular momentum can be transferred to the spacecraft. If the spacecraft's Sun-centered trajectory is deflected during the encounter by 90 degrees, the spacecraft's velocity relative to the Sun is increased by Jupiter's Sun-orbital velocity, about 13 kilometers per second. Jupiter's angular momentum decreases by an immeasurably small amount during the encounter.

Many spacecraft, including the Pioneers, Voyagers, Ulysees, and Cassini/Huygens have used flybys of Jupiter to alter their trajectories. Pioneer 11 and the Voyagers have also performed close flybys of Saturn. One famous probe—Voyager 2—performed close flybys of all of the solar system's giant planets—Jupiter, Saturn, Uranus, and Neptune.

To optimize a gravity-assist maneuver, a spacecraft should pass as close to its target planet as possible, with a low velocity relative to that target

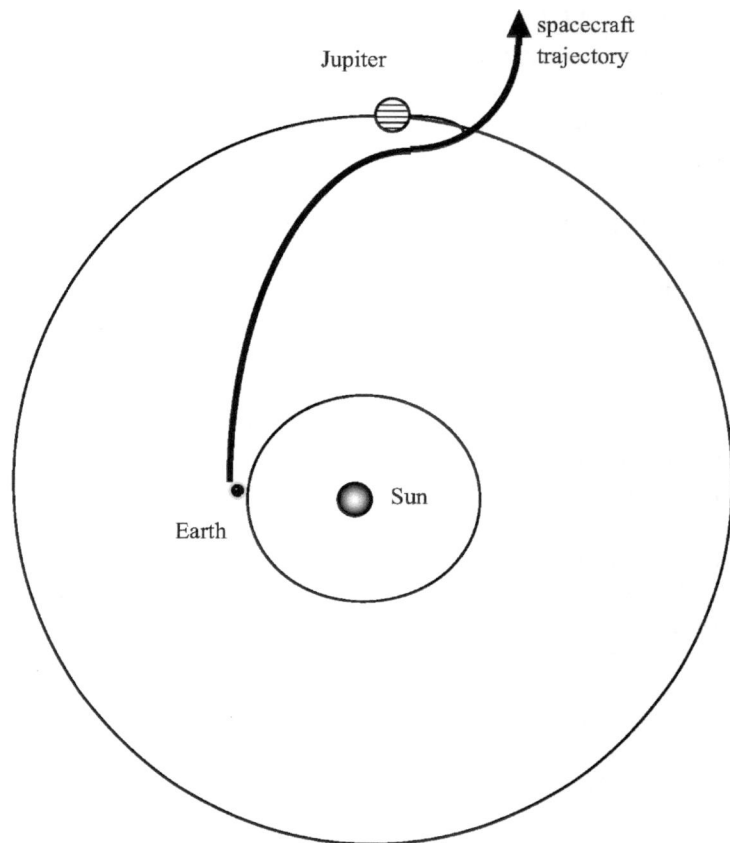

FIGURE 4.1 *A planetary gravity-assist maneuver.*

planet. Voyager 2, the probe that flew past all the solar system's giant worlds, was on a trajectory that was designed to optimize the studies of these giant worlds, not one attempting to achieve a maximum solar-system exit velocity. As a result, its sister ship Voyager 1 is slightly faster, cruising through the fringes of our solar system at about 17 kilometers per second. This is a blistering velocity from a human's point of view—Voyager 1 could traverse the length of Manhattan Island in about one second, but in the cosmic scheme of things, the Voyagers crawl along at a snail's pace. If they were directed toward our Sun's nearest stellar neighbor (which, however, is not the case), the Voyagers would require more than 70,000 years to complete the interstellar transit.

Certain missions have utilized gravity assists from planets smaller than the giants. The Galileo mission to Jupiter was launched using a booster that was too small to reach the giant world directly. As a consequence, this probe spent several years in the inner solar system altering its trajectory with multiple flybys of Venus and the Earth.

Gravity-assist maneuvers to date have been basically unpowered, carefully tailored, flybys of solar-system worlds. But there is another, theoretical, approach that can improve the efficiency of flyby maneuvers.

The gravity field of a planet or star can act as a force multiplier if an impulsive maneuver is performed while a spacecraft is near its closest approach to that celestial object. In other words, a greater velocity increment is possible if the spacecraft is equipped with a rocket that can be fired during the gravity-assist maneuver. Also, firing this rocket deep within a planet's or a star's gravity well produces a greater change in velocity than if the rocket is fired in gravity-free deep space.

Consider, for example, a probe that approaches the Sun within 1.5 million kilometers along a parabolic solar orbit. If a rocket is fired when the spacecraft is closest to the Sun to produce a velocity increase of 2 kilometers per second, the probe will depart the solar system at 41 kilometers per second—about three times the velocity of the Voyager probes!

Spacecraft dynamics using planetary gravitational fields is an evolving discipline that is far from exhausted. Advanced computational techniques have been utilized by mathematicians—notably Edward Belbruno at Princeton University—to analyze a family of low-energy trajectories in the Earth–Moon system. These "weak-stability boundary" trajectories trade increased travel time for reduced thrust requirements. These techniques were first applied in 1991 to transfer the Japanese Hiten probe from Earth orbit to the vicinity of the Moon.

FURTHER READING

Many classical references discuss the technical problems that were addressed to make space travel a reality, including thermal shielding and reentry. Notable among these are Carsbie C. Adams, *Space Flight* (McGraw-Hill, New York, 1958) and Arthur C. Clarke, *The Promise of Space* (Harper & Row, New York, 1968).

Many authors have chronicled the history of the early decades of the Space Age. A wonderful, journalistic treatment of the American manned space program culminating in the Apollo lunar landings is John Barbour's *Footprints on the Moon* (The Associated Press, 1971). A well-researched and beautifully illustrated treatment of the early Russian space effort is Phillip Clark's *The Soviet Manned Space Program* (Salamander Books, New York, 1986).

Space shuttle operations, including thermal-tile design issues, are also treated in multiple sources. One is Robert W. Powers, *Shuttle* (Stackpole Books, Harrisberg, PA, 1979).

Many treatments of space dynamics, including the Hohmann transfer, have been published. If you don't mind the mathematics, you might consult a classic text by Roger R. Bates, Donald D. Meuller, and Jerry E. White entitled *Fundamentals of Astrodynamics* (Dover, New York, 1971).

It's hard to appreciate the fundamentals of unpowered and powered gravity-assist maneuvers without getting into the mathematics. If you have a college-level calculus background and have taken some physics, you might wish to examine *Deep-Space Probes*, 2nd edn (Praxis–Springer, Chichester, UK), which was authored by Gregory L. Matloff in 2005.

This reference also briefly discusses low-energy transfer trajectories in the Earth–Moon system. But for a more comprehensive (and much more mathematical) review of weak-stability-boundary orbital transfers, you could consult Edward Belbruno's *Capture Dynamics and Chaotic Motions in Celestial Mechanics* (Princeton University Press, Princeton, NJ, 2004).

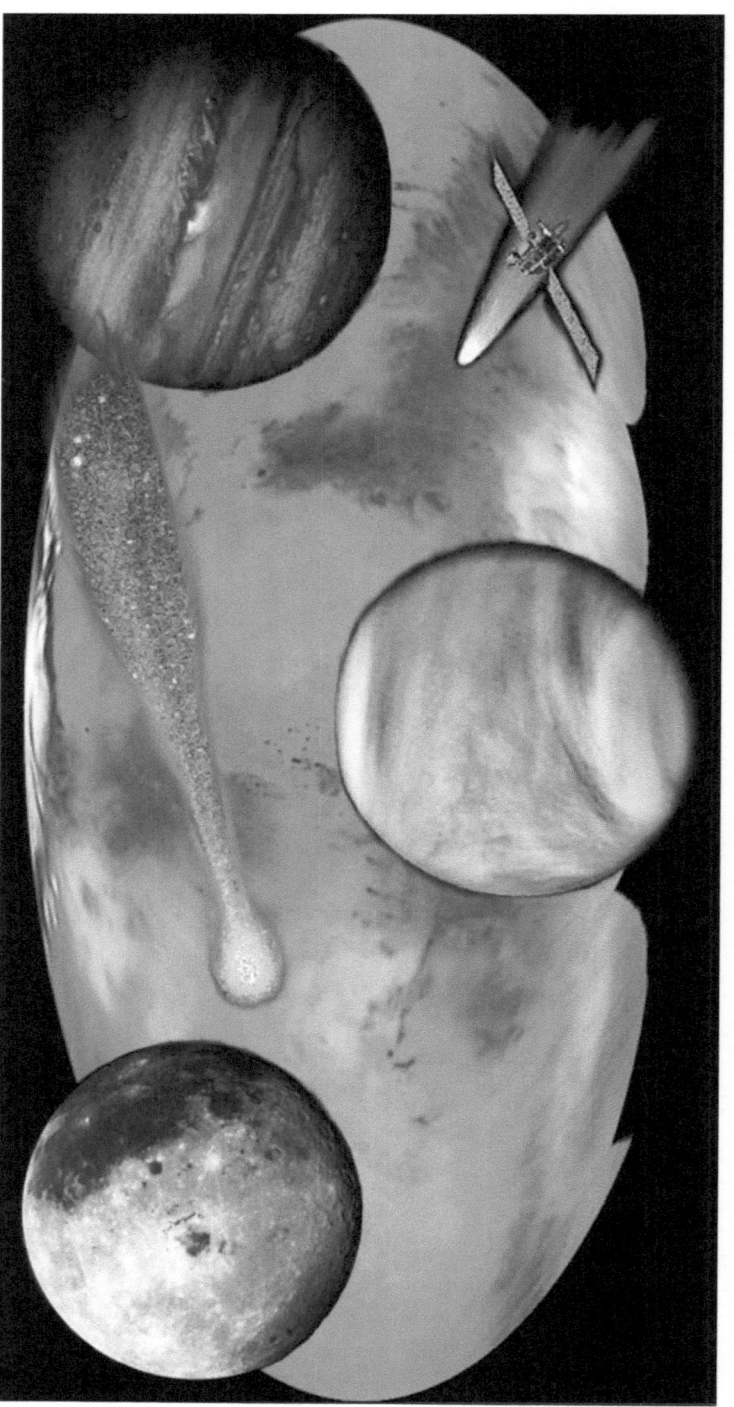

5

PROBES TO THE PLANETS: WHERE WE'VE BEEN ON OUR JOURNEY

The Dog, and the Plough, and the Hunter, and all,
And the star of the sailor, and Mars,
These shone in the sky, and the pail by the wall
Would be half full of water and stars.

Robert Louis Stevenson, from *Escape at Bedtime*

SHORTLY after the beginning of the Space Age, in the late 1950s, rocketeers in the USA and the USSR realized that their primitive, military-derived rockets could escape the gravitational field of the Earth, navigate by the stars to cruise the depths of interplanetary space, and begin the preliminary reconnaissance of the solar system. This was a very heady time for space scientists—war rockets, launched in a superpower competition, would actually be used to study, at close range, properties of distant worlds.

The training ground for these interplanetary Argonauts was our planet's large Moon. At a mean distance of around 384,000 kilometers, Luna would be the easiest destination for space robots. Although the glory would go to the Space Race winning astronauts of the Apollo Program who would walk on our nearest solar-system neighbor, their missions would have been impossible without robot reconnaissance. And much of the scientific harvest of lunar knowledge would be gathered by these precursors to the human exploration teams.

ROBOTS TO THE MOON!

Although they could not sustain the pace to win the race to the Moon, Russian space engineers garnered many of the early milestones in lunar exploration. Russia achieved the first lunar flyby with Luna 1 in 1959. This craft's notable scientific accomplishment was confirmation of the solar wind—the energetic stream of charged particles emanating from the Sun. Even in their failures to achieve Earth-escape velocity, America's Pioneers 1 and 3, both launched in 1958, reached distances sufficient to map the outermost fringes of Earth's Van Allen radiation belts.

On September 12, 1959, Luna 2 achieved a significant geopolitical "first" of little scientific value, when it became the first probe from Earth to impact the Moon. This event was followed in October 1959 with Luna 2, which flew by the Moon and returned the first photographs of the lunar far side. As a result of this mission, some large far-side features have Russian names.

Starting in July 1964, American probes of the Ranger series began their lunar reconnaissance missions. Rangers 7–9 returned excellent, high-resolution photos of the lunar surface as they screamed in toward impact.

America continued robotic lunar exploration with the highly successful Lunar Orbiters, which provided extensive photographic surveys of the lunar topography that would be invaluable to later Apollo astronauts. Successful landings by the Surveyor spacecraft demonstrated in 1966–1968 that Earth ships would not sink into the lunar topsoil. One of these, Surveyor 6, used its rockets to take off and fly a short distance from its initial lunar landing site, demonstrating that the Moon's crust could support the stress of a rocket's exhaust. Parts of another Surveyor were later retrieved by the crew of Apollo 12 and returned to Earth. Analysis revealed that bacterial spores from Earth had survived on board Surveyor for many months in the harsh lunar environment.

After America's victory in the Moon Race, Russia continued a series of scientific Moon missions. In September 1970, Luna 16 performed the first robotic sample-return mission from another celestial body. The quantity of lunar rock and soil returned to Earth by Luna was a tiny fraction of that returned by Apollo astronauts, but this mission paved the way for planned sample-return probes to more distant celestial destinations.

Although America's astronaut-bearing Lunar Roving Vehicles would receive much glory after their deployment from Apollo's Lunar Modules, Russia's Luna 17 succeeded in 1970 in placing the first successful robotic roving vehicle, Lunokhod, on the surface of the Moon. The later Mars-roving robots owe a great deal to this early lunar experiment (Figure 5.1).

FIGURE 5.1 An ancestor to NASA's Mars Pathfinder and Mars Exploration rovers, the Soviet Luna 17 rover explored the lunar surface under remote control. (Courtesy www. mentallandscape.com/ C_CatalogMoon.htm)

After the conclusion of NASA's Apollo Program, there was a long hiatus in both robotic and manned lunar exploration. This was broken by two Soviet successes—the delivery of a second Lunakhod in 1972 and Luna 24's sample return mission in 1976.

Robotic lunar exploration recommenced in 1989 when the NASA Galileo Jupiter probe photographed the Moon during one of its Earth gravitational swing-bys *en route* to its gas giant destination.

A third space power, Japan, got into the lunar exploration act in 1990 with the launch of Hiten, also called Muses-A. This craft is most significant for its translunar trajectory. Instead of flying a Hohmann-transfer orbit requiring three days or so to reach the lunar vicinity, Hiten was injected into a long-duration, but fuel-conservative weak-stability-boundary lunar transfer orbit. Months were required to reach lunar orbit.

In the hope of finding signs of comet-deposited ice in craters near the lunar poles, America launched Clementine and Lunar Prospector during the 1990s. So far, the hunt for lunar water-ice in Sun-shielded craters has not yielded conclusive results.

In 2003, the European Space Agency commenced its program of robotic lunar exploration with the launch of SMART-1. A technology demonstrator, this robot used solar-electric propulsion to spiral from Earth orbit to the vicinity of the Moon. More than a year was required to complete the lunar transfer.

THE LURE OF OUR SISTER PLANET

As the two major space powers began their programs of robotic lunar exploration, it was realized that with some modifications, space probes could survive the five-month journey to Venus, Earth's nearest planetary neighbor. It was necessary to increase rocket power slightly and to harden the probes to survive multi-month exposure to solar radiation. Also, due to the fact that the Earth and Venus separately orbit the Sun, and that poor alignment sometimes makes a journey between them impossible, optimal launch opportunities, or "launch windows" are less frequent than those for missions to the Moon.

Two views of cloud-shrouded Venus were found in the scientific literature in the late 1950s (and in science fiction novels from that era). Might Venus be shrouded in water clouds—and be a benevolent, though warm, habitat for life? Or might the clouds consist instead of carbon dioxide, indicating blistering surface temperatures and an absence of biology? Unfortunately for astrobiologists, the first successful Venus flyby, NASA's Mariner 2, confirmed in 1962 that the high-temperature model was the correct one. Mariner 5 performed more extensive studies of Venus' atmosphere, during its 1967 flyby.

In spite of the far-from-clement conditions near the surface of Venus, Soviet planetary scientists began to concentrate on space probes that would attempt soft landings on Earth's sister world. They finally succeeded with Venera 7, which touched down in August 1970. Alas, contrary to how the Romans imagined her, the planet Venus is no gentle love goddess! No robotic visitor from Earth has survived more than a few hours on the planet's surface. Probes have found that the surface temperature on Venus exceeds 700 Kelvin. The planet's atmosphere is far more massive than that of Earth—and the atmospheric pressure at the surface is about 90 times that at Earth's surface. If that's not bad enough, acidic material constantly rains down on the surface from the planet's lower atmosphere. The conditions on Venus' surface approximate the medieval conception of Hell!

Sadly, Venus will be not become our second home in the solar system. But even though astronauts may never walk upon that world's surface, there is tantalizing science there. And because of Venus' similarity to Earth in terms of size and surface gravity, planetary experts would like to know what strange course of events twisted the evolution of this planet, only 42 million kilometers from the Earth, onto such a divergent path. It's a sobering thought to realize that Earth might not be immune from a Venus-type environmental catastrophe.

Venus exploration has continued at a reasonable pace since the dawn of the Space Age. Many Russian Veneras have landed upon the surface, returned color photos and spectrographic analysis. Two Russian probes—Vega 1 and Vega 2—deployed balloons that survived for two days at an altitude of about 54 kilometers above Venus' surface.

En route to Mercury in 1974, America's Mariner 10 imaged the clouds of Venus in the ultraviolet through infrared spectral bands and surveyed particles and fields in the vicinity of the planet. In 1978, Pioneer Venus 1 and 2 became the first spacecraft to radar map the planet's surface. The radar mapping of Venus continued from several Venera probes and culminated in the magnificent Magellan orbiter, which began its studies in 1989 and demonstrated that Venus is volcanically active. Later, in 1989, NASA's Galileo probe performed imaging studies and spectroscopy of Venus' atmosphere, during its Venus gravity-assist maneuvers *en route* to Jupiter. The robotic exploration of this mystifying world will surely continue.

BLISTERING MERCURY

Because of its proximity to the Sun, tiny Mercury is a difficult destination. Although other missions are being planned, and one is on its way, the only Mercury probe to date is NASA's Mariner 10, which performed three flybys of the planet in 1974 and 1975. Venus gravity-assist maneuvers were essential to project the craft into the near-Sun Mercurian environment.

This craft revealed much about Mercury's orbital dynamics, mass, hyper-thin atmosphere, and (tiny) magnetic field and mapped 45% of the planet's surface. Mercury's heavily cratered terrain appears moonlike in Mariner 10 images.

Mariner 10 was not equipped to conduct radar mapping of Mercury's terrain, but terrestrial radar studies using large radio telescopes have indicated the presence of water ice in Sun-shaded craters near the planet's poles. Probably deposited by comet impacts during Mercury's 4.5-billion-year history, this water may ultimately serve as a resource for human explorers.

This small probe also bears the honor of being the first craft from Earth to operationally apply the principle of solar sailing. (For a detailed discussion of solar sails, refer to Chapter 13.) Owing to the photon-rich environment near Mercury, attitude control fuel was conserved by applying Mariner's solar panels as solar sails to orient the craft.

The Messenger (MErcury, Surface, Space ENvironment, GEochemistry, and Ranging) mission, launched in 2004, will fly by the planet three times before entering orbit in 2011. It will be the first spacecraft to visit Mercury in 30 years! There is no doubt that we will learn a lot more about this oft-forgotten planet over the next few years.

TANTALIZING MARS

No solar-system world excites both the general public and the planetary scientist as the fabled Red Planet. Positioned near the outer edge of the Sun's habitable zone or ecosphere, Mars is the only solar-system world (excluding Earth) on which solar heating alone can result in some non-frozen water. With visible seasonal changes, polar caps that swell in winter and shrink in summer, and a surface gravity about 40% that of the Earth, telescopic astronomers eagerly turned their sights on this enigmatic world.

Noted nineteenth-century astronomers, such as Giovanni Schiaparelli and Percival Lowell, detected linear markings on the planet's surface, at the very limit of visual observation. These observations spawned the legend of the Martian canals, as popularized by early science fiction authors such as H.G. Wells and Edgar Rice Burroughs. Not only might primitive life exist on Mars, but possibly even the remnants of a planet-wide civilization.

And so as the Space Age dawned, mission planners turned their attention toward this fascinating world. At an average distance of 1.52 AU from the Sun, this world presents some challenges to mission planners. Energy-optimized trajectories to Mars are possible about every two years, when Earth and Mars are at their closest. But even during very close planet alignments, such missions would take 6–9 months.

This is a long time for a robotic spacecraft to function in the interplanetary environment. It is interesting to note that less than half of the robots targeting Mars have successfully completed their missions.

First to survive the rigors of launch, trans-Mars injection, the long interplanetary cruise, and the appetite of the "Great Galactic Ghoul" that seems to claim so many Mars-bound craft, was America's Mariner 4. Launched in 1964, this craft flew by the Red Planet and photographed less than 1% of its surface. Alas, there were craters aplenty—but no canals! In 1969, Mariners 6 and 7 photographed Mars during their flybys and conducted atmospheric studies, confirming the findings of their earlier sistership. The canals were an optical illusion; technological extraterrestrials had not evolved on Mars.

But flybys could survey only a fraction of the planet's surface and it was necessary to conduct a long-term mapping survey from Mars-orbit. Mariner 9, the first successful Mars orbiting probe, was launched in 1971. Not only did this craft photograph essentially the entire surface of the Red Planet; it also provided high-resolution photos of Mars' tiny satellites Deimos and Phobos.

Mariner 9's orbit around Mars was quite eccentric, with a low point of about 500 kilometers and a high point of about 16,000 kilometers above Mars' surface. This ninth (and final) Mars Mariner is one of the most successful interplanetary missions to date. During its one-year operational lifetime, this 570-kilogram robot snapped more than 7,000 photos of Mars and its satellites. Planetary scientists studied Martian dust storms at close range and learned that wind-blown dust was a major conributor to seasonal changes on the Red Planet.

But some of Mariner 9's photos revealed a great diversity in Martian terrain. There were vast dormant volcanoes that dwarf Mount Everest, sinuous markings that resemble extinct river valleys. One enormous rift valley was charted—Valle Marineris is far larger than the Grand Canyon and Olduvai Gorge—it may have been visible to nineteenth-century terrestrial observers and been mistaken as a canal.

Visionaries planning future human habitation of Mars were especially inspired by at least one finding of Mariner 9. Clouds were observed over the planet's polar caps, and it was suggested that these were perhaps composed of water. Combined with moderately clement equatorial temperature readings during daylight hours in Martian summer, this data

FIGURE 5.2 The contributions made by the two Viking Landers are often overlooked during our present-day renaissance of Mars planetary exploration. (Courtesy NASA)

indicated that Mars might not be totally inhospitable to properly protected terrestrial life.

Mariner 9's observations of the thin Martian atmosphere were instrumental in planning the next US missions to the planet—probes that would touch down on the planet's surface and search for signs of life.

Suddenly, Mars was no longer a dull, moonlike world. For a while it seemed not impossible that at least simple life may be found there.

In an attempt to settle this issue, Vikings 1 and 2 arrived in Mars orbit in 1976. These craft each consisted of an orbiter plus a lander. For years, planetary scientists were treated to a plethora of photographs from both surface and orbital locations (Figure 5.2).

A series of photographs from the Viking orbiters revealed a mysterious structure in a region called Cydonia that resembled a sculpted human head. This "Great Face of Cydonia" excited tremendous public interest in Mars exploration, even after higher resolution images from a later probe revealed that it is almost certainly a dust-covered dune in the Martian desert.

Although the life-detection experiments were inconclusive, the Viking landers surveyed Martian geology and operated as weather stations. The north polar cap contained frozen water, not only dry ice (frozen carbon dioxide) as many planetary scientists had suspected.

The Viking landers discovered that organic compounds—molecules based on carbon—were absent from the planet's surface layers. Perhaps Mars' atmosphere was the culprit. Much thinner than Earth's atmosphere, Mars' atmosphere admitted high-energy solar electromagnetic radiation that might destroy complex molecules like those required for known types of life to exist.

But what about the subsurface layers? Viking revealed the presence of permanently frozen subsurface water—or permafrost. Perhaps there were regions where underground lakes existed. Perhaps ancient Martian life had evolved on the planet's surface during the early evolution of the solar system and migrated undergound as conditions worsened.

Interest in Martian life reached near-fever pitch in the late 1990s after analysis of a meteorite found in the Allen Hills of Antarctica. The meteorite, called ALH84001 (Allen Hills, 1984 #001) was determined to have originated on Mars and revealed microscopic forms suggestive to many of fossilized life. Since the science of life detection (on another world!) is not well developed, NASA adopted an exploration strategy best described, as "follow the water." All known forms of life require some sort of water to survive. If life exists or existed on Mars, it is more likely to be found near water. To this end, a series of orbital missions and landed rovers began their exploration of Mars, searching for the elixir of life. These next-generation planetary explorers would land in a new fashion, with cushioned beach-ball-like balloons, and deploy rovers to explore regions distant from the relatively smooth landing sites—regions with very interesting geology where, just possibly, extinct or existing life might be found.

A NASA probe, the Mars Global Surveyor, was launched toward the Red Planet in 1996. As well as continuing the surface and atmosphere reconnaissance from orbit, this craft had the capacity to communicate with landed probes and could serve as a data relay to Earth.

The first of the new rovers was a NASA technology demonstrator called Pathfinder that carried the experimental rover Sojourner. This highly successful test of a new landing technique was flown in 1997, and not only demonstrated the utility of airbag-cushioned Martian descent but also showed that rovers on the Martian surface could operate far from their mother ships. This mission also demonstrated the public's interest in further Mars exploration. The Pathfinder Website was extremely popular and gathered a record number of hits (for a government website!) during the surface phase of the mission.

Following the failed 1998 Mars Climate Orbiter and Mars Polar Lander missions, and the successful Mars Odyssey (2001) orbiter, NASA launched two more sophisticated rovers—Spirit and Opportunity. Arriving on Mars in early 2004, these rovers have ventured kilometers from their landing sites. They have returned thousands of images of Martian terrain features and have conducted microscopic analysis of geological features, some of which have been gathered by subsurface drilling. Some of the geological findings point to the almost certain ancient existence of widespread Martian oceans, and lakes.

While suffering more than their fair share of lost missions, the Russians (formerly the Soviet Union) also contributed a great deal to humanity's knowledge of Mars. The Mars 2 and Mars 3 orbiters (1971) returned numerous images of the planet's surface, allowing the construction of early topographic maps. A follow-on series of orbiters and landers, Mars 4, 5, 6 and 7 were largely unsuccessful and returned only a very limited data set. Two missions were launched to Mars' moon Phobos in 1998. The first was lost in transit, but the second successfully orbited Phobos before communications were lost.

The Japanese Mars orbiter mission, Nozomi (1998), Japan's first attempt to send a probe to Mars, also failed to reach the planet. And the Europeans' Mars Express (2003) is busily sending data back from Mars orbit, though its Beagle 2 Lander was lost. One significant bit of data from Mars Express was its discovery of atmospheric methane—which greatly increases the possibility of there being surviving, subsurface life some-where on the planet.

Exploring another world is challenging indeed!

DISTANT GIANTS

Shortly after the dawn of the Space Age, it was realized that a planetary alignment would occur in the 1970s that would not be repeated for 179 years. A properly aimed spacecraft could fly past each of the gas giant planets in our solar system—Jupiter, Saturn, Uranus, and Neptune—using gravity assists to increase spacecraft orbital energy and deflect the trajectory at each encounter in preparation for the next.

NASA planned four missions to accomplish this task. Before the end of the Pioneers 10 and 11 and Voyagers 1 and 2 missions, humanity had photographed at close range each of the giants and many of their satellites and gathered valuable information on planetary and interplanetary environments. After the final planetary flyby, as the craft all headed into the interstellar abyss, one turned its camera sunward and photographed its distant world of birth, swimming in a band of scattered sunlight like a "pale blue dot."

Mission planners realized that there was a chance, however slight, that these tiny probes could be intercepted by spacefaring extraterrestrials as they cruised the galactic wilderness. Therefore, each was equipped with a message plaque—a "message in a bottle" dropped by us into the galactic ocean.

Pioneer 10 entered Jovian space in December 1973. After conducting a preliminary survey of the giant planet's cloudbanks, rings, satellites, and radiation belts, Pioneer used Jupiter's gravitation field to alter its trajectory and propel it toward the stars.

One year later, Pioneer 11 encountered giant Jupiter. This time, however, the trajectory was altered so that the probe would flyby Saturn on its way out of the solar system.

The preliminary findings and images from the Pioneers gave mission planners the opportunity to meticulously plan the missions of the much more capable Voyagers. Voyager 1 encountered Jupiter in March 1979 and flew by Saturn in November 1980. The final grand-tour probe, Voyager 2, also flew by Jupiter and Saturn. It then encountered Uranus in 1986 and Neptune in 1989.

These craft should be considered extrasolar rather than interstellar probes. They will leave the region of space influence of the Sun, but the spacecraft are not intended or designed for travel to another star. The fastest of them, Voyager 1, is cruising through the interstellar night at about 17 kilometers per second or 3.5 AU per year. If it were directed toward our Sun's nearest stellar neighbor (which it is not), it would get there in approximately 70,000 years.

The photographic survey of the outer planets conducted by these four craft has been invaluable, as have the particles and fields measurements of the outer interplanetary and near-interstellar environments. Of most significance to future mission planners are the imaging of Jupiter's large satellite Europa, which indicated the presence of a water ocean beneath a thick ice crust. Another large Jovian satellite, Io, was found to be in a state of perpetual volcanic activity. Studies of Titan, the largest satellite of Saturn, revealed that this object has a dense atmosphere and surface conditions that might resemble those of an early Earth.

To further study these enigmatic satellites and the environments of their parent worlds, two ambitious follow-on missions were planned. Galileo would be directed to enter orbit around Jupiter, while Cassini would become a satellite of Saturn and deposit the Huygens probe on Titan's surface.

Galileo was originally scheduled to be launched by a Centaur high-energy upper stage deployed in Earth orbit from a space shuttle. But after the Challenger accident of 1986, it was decided that the use of Centaur might pose risks to a shuttle crew. As a less energetic rocket was substituted, multiple passes of Earth and Venus were therefore required before Galileo could be injected into a trans-Jupiter trajectory.

During its sojourn in the inner solar system and its flight to the giant planet, Galileo imaged Earth and Moon from deep space, as well as asteroids Gaspra and Ida. It was discovered that Ida has a small satellite asteroid that has been dubbed Dactyl.

Finally arriving at Jupiter in December 1995, Galileo entered orbit around the giant planet. Before it adjusted its interplanetary trajectory, Galileo dispatched a subprobe which entered Jupiter's atmosphere and

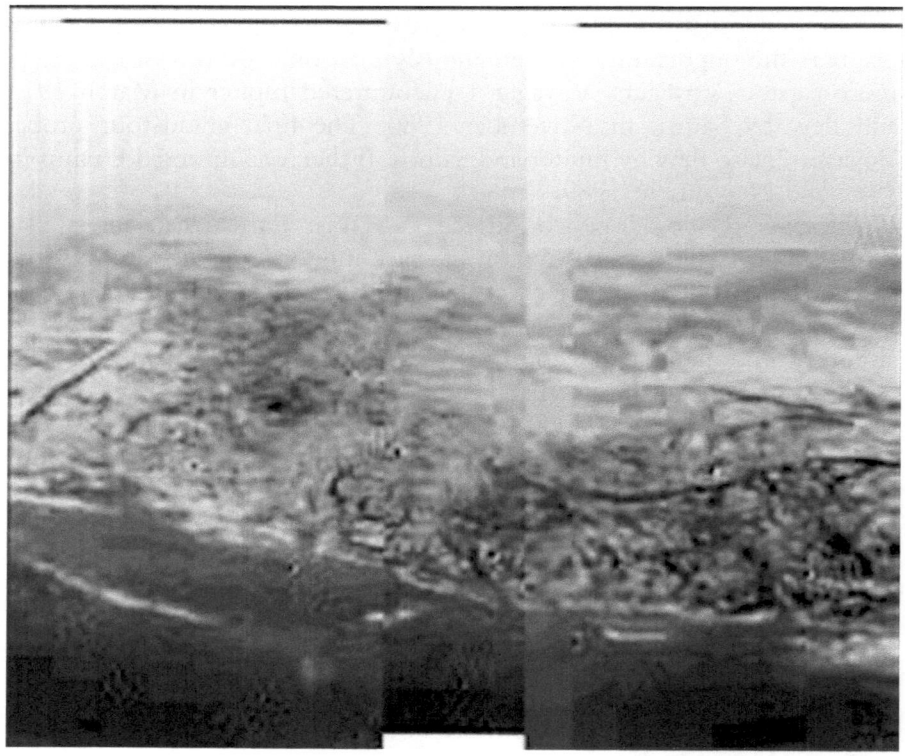

FIGURE 5.3 Yet another New World was seen as the Huygens Probe descended to the surface of Saturn's moon Titan. (Courtesy NASA)

radioed data from a depth of a few hundred kilometers below the visible cloud deck.

During its years of operation, Galileo surveyed Jupiter's large and small satellites, its rings, and accurately mapped the radiation environment of this giant world. Its most significant results in the long term may be close-up photographs of Europa, revealing that there are apparent cracks in the ice sheet. If liquid water exists near the surface of this satellite, future probes may find life there.

On July 1, 2004, the Cassini spacecraft entered orbit around Saturn and relayed to Earth many beautiful pictures of that planet and its ring system. A spectacular accomplishment of the mission was the successful operation of the Huygens Titan lander. This subprobe survived a perilous descent to the surface of Saturn's moon, Titan, which has a hydrocarbon atmosphere about as dense as the Earth's. Although most terrestrial life forms would quickly expire in Titan's harsh environment, some form of life may be possible in liquid methane "springs" and "lakes" on this remote and very distant surface (Figure 5.3).

SMALL WORLDS

Although the robotic exploration of the planets has gathered much attention, missions have also been conducted to smaller solar-system bodies—comets and asteroids. Space scientists and mission planners are interested in these diminutive denizens of the solar system for several reasons. First, they contain vital clues about the origin and evolution of the solar system. Second, they contain materials that may provide a resource base for an expanding human presence in space. Finally, these objects occasionally collide with Earth—with devastating consequences. Only *in situ* measurements can help us to devise the best methods to mitigate these catastrophes.

In 1986, one of the most famous of these objects paid a periodic visit to the inner solar system. During the 1986 apparition of Halley's Comet, it was visited by a host of spacecraft launched by Europe, Japan, and the USSR. Europe's Giotto entered the comet's coma and rapidly cruised past its nucleus, transmitting magnificent images and miraculously surviving the perilous passage to encounter a second comet sometime later.

In 1996, NASA launched NEAR (Near-Earth Asteroid Rendezvous) toward asteroids Mathilde and Eros. After flying past Mathilde, NEAR continued on toward Eros. It orbited this near-Earth asteroid and soft-landed on the surface at the conclusion of its mission (Figure 5.4).

FIGURE 5.4 Asteroid Mathilde, as photographed by the NEAR spacecraft. (Courtesy NASA)

Building upon NEAR's success, NASA launched Deep Space 1 in 1998. Conceived as a technology demonstrator of solar-electric propulsion (SEP) and other technologies, Deep Space 1 imaged asteroids and a comet on its journey through the inner solar system.

The following year, NASA's stardust probe rocketed toward a rendezvous with Comet 81P Wild 2, in the first attempt to return comet samples to Earth. Stardust, and its celestially harvested samples, returned to Earth in 2006. NASA's most recent comet venture, Deep Impact, dispatched a subprobe to impact Comet P/Temple 1 on July 2005. For the first time, space scientists gathered information about a comet's interior.

Europe and Japan have not been idle in the exploration of small solar-system objects. Japan's Hayabusa was launched in 2003. Equipped with solar-electric propulsion, this craft was directed toward asteroid 25143 Itokawa. Samples from this asteroid are to be collected and returned to Earth in 2007. In March 2004, Europe's Rosetta was launched toward a 2014 rendezvous with Comet 67/P Churyumov–Gerasimenko. If all goes according to plan, Rosetta will orbit the comet and deposit a lander on its surface.

Future scientific missions to small solar-system objects are planned. At least one of them may be privately funded.

FURTHER READING

A 1957–1998 chronology of robotic solar-system space missions is included in Katharina Lodders' and Bruce Fegley Jr's *The Planetary Scientist's Companion* (Oxford University Press, New York, 1998).

Many popular treatments of Mars exploration have been published. One beautiful popular edition is: Martin Caidin, Jay Barbree and Susan Wright, *Destination Mars* (Penguin Putnam, New York, 1997).

European robotic space missions are summarized in a number of sources. One is: Giancarlo Genta and Michael Rycroft, *Space: The Final Frontier* (Cambridge University Press, New York, 2003). Aspects of the Galileo and Cassini/Huygens missions are also discussed in this source.

One source of current information regarding the status and accomplishments of robotic space missions is *Spaceflight*, a popular periodical published by the British Interplanetary Society. For a review of the early days of the Spirit/Opportunity missions to Mars, consult an article in the March 2005 issue of that magazine authored by Philip Corneille, entitled "Roving the Red Planet," *Spaceflight*, vol. 47, pp. 102–106.

Recent Cassini/Huygens results are reported by Clive Simpson, in articles entitled "The New Lord of the Rings," *Spaceflight*, vol. 46, pp. 353–355 (2004) and "Titan—A Glimpse into the Unknown," *Spaceflight*, vol. 47, pp. 94-99 (2005).

6

PROBES TO THE STARS: CONCEPT STUDIES

Thou fair-hair'd angel of the evening,
Now, whilst the sun rests on the mountains, light
Thy bright torch of love; thy radiant crown
Put on, and smile upon our evening bed!

William Blake, from *To the Evening Star*

FOR countless generations, humans have peered up into the evening sky and wondered about those strange lights yonder. Could we touch them; could we reach them? For most of human history, such speculations were the province of the poet or the writer of visionary fiction.

Shortly after the development of the nuclear bomb, however, things began to change. The stars might shine very distantly, seemingly unreachable across a vast abyss of space, but human ingenuity might be capable of bridging the gap, even if the journeys might span millennia.

One breakthrough on the road to the stars occurred in an unexpected manner. An American physicist, Theodore Taylor, performed an experiment at a nuclear test site demonstrating that certain objects could survive the hellish environment of a nearby nuclear detonation. An aluminum sphere, coated with graphite, "surfed" on a nearby nuclear blast with little damage.

The trick was ablation. The energies of the nuclear detonation evaporated or boiled-off graphite molecules at high velocities, shielding the aluminum underlayer from the blast.

This was the birth of Project Orion, a concept for a piloted spacecraft that would be propelled by nuclear explosives. Each nuclear explosion would give the spacecraft a "kick," propelling it to higher and higher speeds. In 1968 when the top-secret program began under the auspices of the US Department of Defense, Orion was conceived as an interceptor of Soviet spacecraft threatening the USA. Later, when NASA became involved, Orion's interplanetary possibilities were investigated.

Initially, Orion was to be propelled by nuclear fission charges, each of which would have an explosive yield between 0.01 and 10 kilotons of TNT. These could be "shaped" to optimize momentum transfer to the spacecraft. The first-generation Orion would have an effective exhaust velocity of 20–60 kilometers per second, which might be uprated to 100–200 kilometers per second.

After ignition, nuclear debris would interact with the craft's ablative-covered "pusher plate." A series of enormous shock absorbers would smooth the ride for Orion's occupants.

Owing to the Nuclear Test Ban Treaty, a full-scale Orion spaceship was never launched. But a small model propelled by chemical charges successfully lifted off, flew for about 60 meters, returned to Earth by parachute, and is now on display in the Smithsonian Air and Space Museum in Washington, DC.

Another factor in Orion's cancellation was the successful testing of the large chemical rockets of the Saturn series that would ultimately carry American astronauts to the Moon. Exotic nuclear drives would not be necessary to defeat the Soviets in the Space Race.

Before the project's demise, engineers demonstrated that Orion could be used as a Saturn upper stage, thereby eliminating the problem of nuclear fallout. Physicist Freeman Dyson was impressed with the concept's potential for interplanetary flight. If development had proceeded beyond 1965, astronauts might have reached Mars by now. Orion's documentation and potential might have disappeared into a classified file and been hidden forever from the light of day, but Professor Dyson saved the day, and showed that interstellar travel was not forever forbidden.

INTERSTELLAR H-BOMBS

Dyson choose to discuss Orion's interstellar potential in a scientific periodical with the widest possible audience. He published his epochal article, "Interstellar Transport," in the October 1968 issue of *Physics Today*.

In that article, Dyson first presented the Orion concept and then asked the reader to consider a massive spacecraft that would be constructed in orbit. Dimensions of such an interstellar Orion would be in the kilometer range (Figure 6.1). Using basic principles and unclassified nuclear data, he demonstrated that the USA and the USSR thermonuclear arsenals could be utilized to expand humanity, not destroy it.

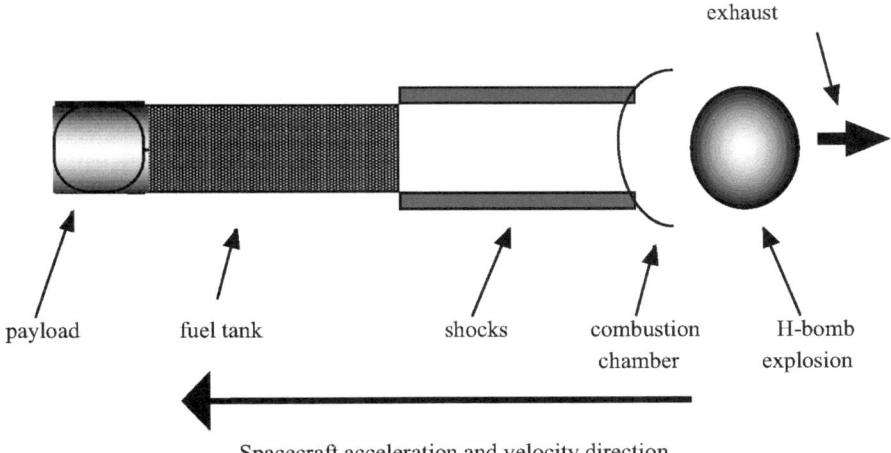

FIGURE 6.1 An Orion H-bomb-propelled starship.

Using Orions propelled by thermonuclear detonations, thousands of human migrants could depart for the nearest stars. Without divulging classified information that would undoubtedly result in a long jail sentence, Dyson could not pin down the performance of his huge ships, but he was able to demonstrate that the crossing to Alpha Centauri (the nearest star system at about 40 trillion kilometers or 4.3 light-years from ours) would take 130–1,300 years.

What a wonderful thing to do with all the world's H-bombs! Although the economics of constructing the huge ship and transporting it and its fuel charges into orbit is daunting, Dyson projected that the world might be up to this task in the twenty-second century.

This was a heady moment for interstellar enthusiasts. But analysts under-appreciated the political difficulty of convincing powerful governments to productively dispose of their dangerous toys. Also, some realized that the mental health of a starship's crew would not be favored by a one-megaton nuclear blast igniting less than a kilometer away every few seconds, no matter how effective the shielding!

In the early 1970s, the British Interplanetary Society took up the challenge and initiated their own starship study. This Orion-offspring was christened Project Daedalus, and, in hindsight, it adopted the philosophy of "living off the land," scavenging its fuel from a multitude of potential off-planet sources.

A SANITIZED ORION

Although nuclear-fusion-propelled starships could certainly carry people, Daedalus was originally conceived as a robotic star probe. The selected target was Barnard's Star, which is approximately 6 light-years from the solar system and was (erroneously) suspected of possessing a planetary system in the 1970s.

Instead of thermonuclear-bomb propulsion, inertial fusion was selected for the Daedalus motor. An inertial fusion reactor works by using electron beams or lasers to compress and ignite fusion-fuel pellets (Figure 6.2).

Contemporary attempts to tame thermonuclear fusion usually react two heavy isotopes of hydrogen—deuterium and tritium. Although fusion reactors utilizing these fuels are approaching technological feasibility, there is a major problem, at least from the point of view of starship propulsion.

As well as lots of energy, the byproducts of deuterium–tritium fusion include high-energy neutrons. A terrestrial fusion reactor would be equipped with massive walls to protect the surrounding population from this radiation. But mass is at a premium in spaceflight.

The Daedalus study team therefore elected to propose a different thermonuclear reaction—that of deuterium and helium-3. This reaction, which produces far fewer neutrons, is also approaching technological feasibility.

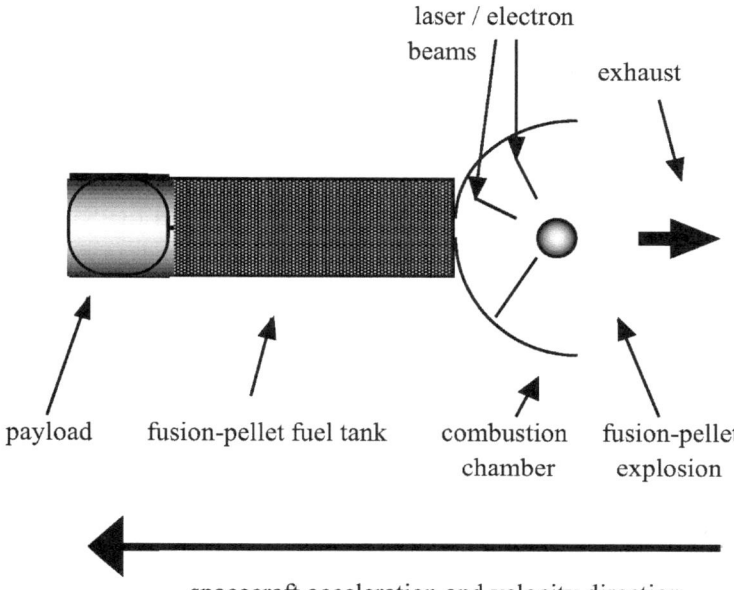

laser / electron beams

exhaust

payload   fusion-pellet fuel tank   combustion chamber   fusion-pellet explosion

spacecraft acceleration and velocity direction

FIGURE 6.2 The Daedalus inertial-fusion starship.

Helium-3, a low-mass helium isotope, is unfortunately quite rare on Earth. A civilization conducting Daedalus-type starflight would have to electromagnetically tap the solar wind, strip mine the Moon, or obtain this resource from the atmospheres of the giant planets. The Daedalus team proposed that robot balloons suspended in Jupiter's atmosphere could accomplish this task.

An additional problem hindering the development of Daedalus-class (or Orion-class) interstellar vehicles is that of scale. Whether the payload is a 100-million-kilogram interstellar colony ship, a 500,000-kilogram robotic planetary exploration probe, or a 100-kilogram miniaturized interstellar medium probe, the mass of the propulsion system is the same—and a fusion-pulse propulsion system is enormous. There just does not seem to be a way of scaling the development of such a rocket, and the capital outlay would be enormous.

Daedalus has enormous potential—it could theoretically achieve an exhaust velocity of about 10,000 kilometers per second (about 3% the speed of light). Robotic probes could be accelerated to 12% of light speed if directed to fly through nearby planetary systems; larger colony ships could certainly reach at least 1% of light speed. But because of propulsion-system scaling and fuel-collection economics, Daedalus-class vehicles must be postponed to the time when humanity has created a resource-intense solar-system-wide space infrastructure. Sadly, the Daedalus team began to consider, and reject, some alternatives.

SOME WONDERFUL IMPROBABILITIES

One possibility considered by the Daedalus researchers was the interstellar ramjet (Figure 6.3). Conceived in 1960 by the American physicist Robert Bussard, an ideal ramjet would use an electromagnetic (EM) field to scoop up protons from the interstellar medium and "burn" them in a fusion reactor to obtain helium and energy, thus allowing the ramjet to "live off the land" in interstellar space. Although it is the only physically possible interstellar propulsion scheme capable of reaching arbitrarily relativistic velocities, there are very major technological problems associated with the proton-fusing ramjet.

First is the comparatively minor point that most EM field configurations tend to reflect interstellar ions rather than collect them, thereby functioning as excellent drag brakes. A much more significant flaw is the

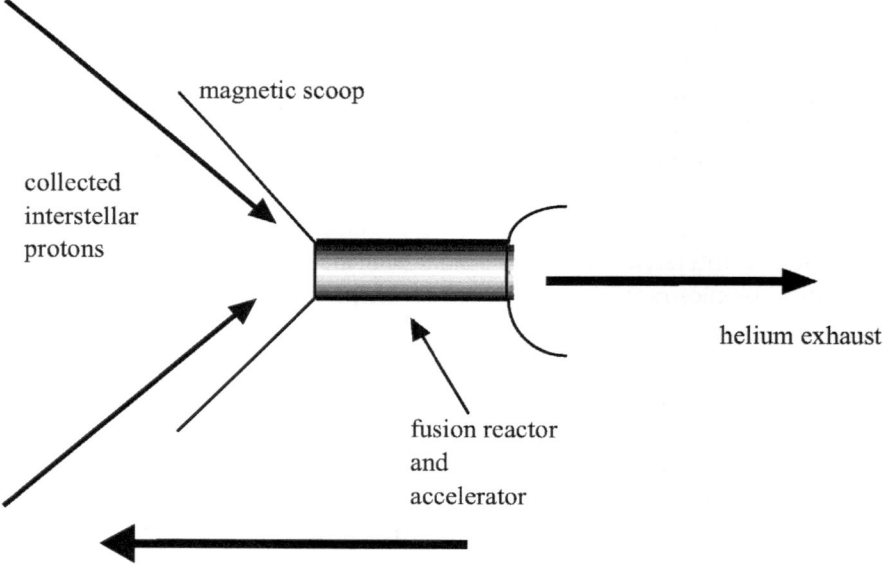

FIGURE 6.3 A proton-fusing interstellar ramjet.

fact that proton-fusion (the reaction that powers the Sun and most stars) may forever be beyond our technological reach.

In a nuclear fusion or fission reaction, less than 1% of the reactant mass is converted into energy. But what if we could increase this energy-conversion efficiency dramatically?

In theory, this can be done by combining fuel atoms with antimatter. The matter/antimatter reaction, made famous by *Star Trek*, converts all of the reactant mass into energy. A matter/antimatter rocket can theoretically achieve relativistic velocities without enormous mass ratios. For more information about antimatter propulsion, please refer to Chapter 20.

But (fortunately for terrestrial life) antimatter is exceedingly rare in the present-day cosmos. Producing it in specialized nuclear accelerators, or "antimatter factories," is enormously expensive. Perhaps it's a good thing in this terrorist-plagued world that production of even a gram of this substance would break the world's economy.

Another antimatter problem is long-term storage. This exceedingly volatile substance combines very rapidly and explosively with normal matter. If the electromagnetic containment field ever fluctuated (as it did occasionally in *Star Trek*), the starship might be momentarily visible across the galaxy as an exploding star.

The final exotic possibility to be considered here is space–time warping. A favorite of the science fiction author and movie producer, the space warp would locally alter the fabric of space–time so that a spacecraft could take a superluminal shortcut and voyage between stars on short-duration journeys.

Although general relativity theory indicates that space–time can be warped by sufficiently high mass densities, electromagnetic fields, and angular momentum (spin), it is not easy to put these principles into practice unless you can collapse a star's mass—perhaps several of them. Some studies indicate that creation of a non-gravitational "warp-bubble" around a spacecraft might require more energy than exists in the universe.

Even if we learn how to create such a "singularity" in space–time, there is another small problem. No one knows how to control the trajectory (if that is the correct word) of a spacecraft traversing some higher dimensional hyperspace.

TAU: NASA's First Interstellar Probe Study

As the Daedalus study progressed, a number of other space agencies and organizations began to express an interest in robotic interstellar travel.

Foremost among these was the NASA Jet Propulsion Laboratory (JPL) in Pasadena, California.

In 1976, mission planners realized that Pioneers 10 and 11 and Voyagers 1 and 2, although they would escape the solar system, would not survive long enough to survey the interstellar environment more that 100 AU or so from the Sun. To obtain useful data regarding the near-interstellar environment, chemical rockets and planetary gravity assists had severe limitations. A radically new propulsion system was required.

At JPL, a number of engineers and scientists banded together to consider the type of interstellar mission that could be reasonably launched in the time frame 2000–2050. It was determined that the best that could be hoped for was a mission to a Thousand Astronomical Units (TAU). TAU would carry a suite of scientific instruments designed to measure electromagnetic fields and particle densities at and beyond the fringe of the solar system.

Optical-imaging equipment could also be included. By correlating photographs of star fields between TAU's telescopes and terrestrial instruments, it was demonstrated that accurate stellar distances could be determined for very distant stars.

TAU would be a large, fast spacecraft. It would have to attain 100 kilometers per second to reach 1,000 AU from the Sun within a human working lifetime. Only two near-term propulsion systems were deemed to be technologically and politically feasible. One was the nuclear-electric rocket (Chapter 3).

An onboard fission reactor would be utilized to ionize (cesium, argon, or mercury) fuel atoms and to accelerate these atoms to a velocity of 100 kilometers per second. One reason for TAU's size was fuel require-ments—about two-thirds of the launch mass would be ion fuel.

Because this fuel would be exhausted over a period of years or decades, electric-engine reliability in the deep-space environment must be very high. TAU engineers therefore investigated a number of alternatives to the nuclear-electric rocket. Interestingly, their findings were identical to those of Daedalus investigators looking for an alternative to nuclear-pulse propulsion for interstellar colonization missions.

STAR SAILING

If we construct in space the thinnest, most highly reflective and temperature tolerant possible solar sail, attach it to its payload with the

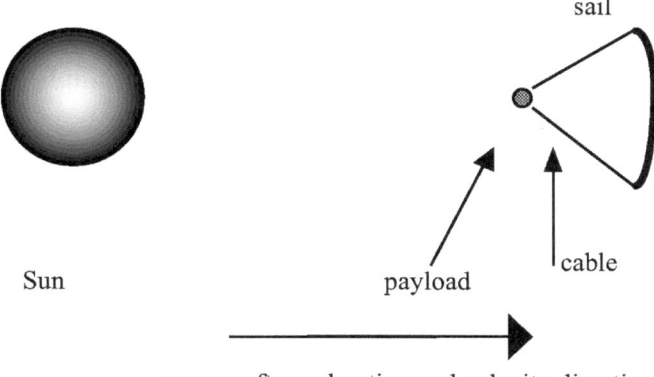

sail

Sun   payload   cable

spacecraft acceleration and velocity direction

FIGURE 6.4 An interstellar solar sail.

strongest possible cables, we can use this device to conduct interstellar missions. (As described in Chapter 13, solar sails are propelled by the force of solar photons reflected from the sail.) It's necessary to unfurl the sail as close to the Sun as possible, from an initially parabolic or hyperbolic solar orbit, to achieve a high interstellar velocity (Figure 6.4). But interstellar solar sails are slow—the best possible crossing time to the Alpha Centauri system is about 1,000 years. This is the approximate performance of a star ark propelled by thermonuclear-pulse propulsion.

Now if we downgrade our sail and structure to the best possible Earth-launched configuration projected for the first few decades of the twenty-first century, performance is degraded. But it is not impossible that such a reasonably near-term sail could approximate the performance of the TAU nuclear-electric rocket.

Extrasolar mission planners began around 1990 to consider the solar sail. They were attracted by the fact that utilizing the sail rather than a nuclear rocket would be far easier from a sociopolitical standpoint.

Also, unlike the nuclear drives, the solar sail is scalable. Early extrasolar missions with payloads in the 10- to 30-kilogram range utilizing sails less that 200 meters in radius could yield information valuable to the designers of true interstellar missions utilizing sails 100 kilometers or larger in radius.

In 1992, a team of European and American engineers and scientists began to consider a sail mission to the Sun's gravitational focus. Gravitational lenses, many of which have been discovered in intergalactic space, occur when a massive celestial object is between a more distant object and the observer.

The gravitational field of the closer object greatly amplifies the light from the more distant object. The image of the occulted object will be distorted: some intergalactic gravitational lenses have multiple images of the more distant object, appearing to form a cross in space.

Although the amplified beam from such an "Einstein cross" continues to infinity, the focus begins a distance from the closer object. If one disregards solar coronal effects, which may push the gravitational focus farther out, the Sun's gravitational focus begins about 550 AU from the Sun.

If astronomers wish to obtain greatly amplified images of a celestial object, say the super-massive black hole at the center of the galaxy, they might launch "ASTROsail" in the direction opposite the Milky Way's center. After passing the 550 AU inner solar-gravitational focus, greatly amplified images would be obtained and relayed back to Earth. If radio astronomers succeed in their searches for radio transmissions from extraterrestrial intelligence (SETI), a SETI sail operating beyond 550 AU could provide excellent data on ET's home solar system, if that system were occulted by the Sun.

But even with payload micro- (or nano-) miniaturization, and even assuming reasonable advances in sail technology and supplemental use of giant-planet gravity assists, the duration of such a mission would be many decades. A nearer extrasolar target than the Sun's gravity focus would be desirable, at least for a preliminary technology-demonstration mission. If the focal probes could be considered as interstellar precursor missions, a less-demanding extrasolar sail might be considered an interstellar-precursor precursor!

Such a concept was the European Aurora study of the mid-1990s. Instead of aiming for the Sun's gravity focus, Aurora was to target the much nearer heliopause.

The interaction between the Sun and its surrounding galaxy is complex. When the radioactive batteries on board the Voyagers finally run out of juice around the year 2020, these two small robots will be perhaps 150 AU from the Sun. If they could hold out another few decades, scientists could receive particles and fields data from the Sun's galactic vicinity.

The particles/fields boundary between the Sun's influence and that of the galaxy is called the "heliopause." Located at approximately 200 AU from the Sun, the heliopause was to be the goal of the conceptual Aurora probe.

This craft would be equipped with a microns-thin solar sail capable of withstanding a pass to within 0.3 AU from the Sun. Owing to the difficulty of launching a very thin solar sail from Earth and unfurling it in space, some Aurora researchers proposed that the sail could be deposited on a thicker plastic substrate that would be sensitive to ultraviolet radiation and would quickly sublimate when exposed to sunlight.

Other mass-reducing innovations for the proposed 250-meter Aurora sail included inflatable sail beams and supports. The total mass of the Aurora spacecraft was estimated to be 150 kilograms. It would likely be placed in an Earth-escape trajectory by a chemical upper-stage rocket prior to unfurlment.

Because of its low mass and close solar approach, Aurora was projected to exit the solar system with a velocity in excess of 12 AU per year. Several times faster than Voyager, Aurora would pass the heliopause less than two decades after launch and might well survive to transmit data from the Sun's gravitational focus.

As is the case with many studies, Aurora ran its course and was never launched. However, many of the technological innovations projected for this craft have been folded into the NASA's hoped-for Interstellar Probe project—a more advanced sailcraft that NASA hopes to launch toward the heliopause around 2020. This extrasolar probe concept is further described in Chapter 8.

FURTHER READING

A popular treatment of the interstellar probe concept is Paul Gilster's *Centauri Dreams* (Copernicus, New York, 2004). At a somewhat higher mathematical level and a bit more dated are Eugene Mallove's and Gregory Matloff's *The Starflight Handbook* (Wiley, New York, 1989) and John H. Mauldin's *Prospects for Interstellar Travel* (Univelt, San Diego, CA, 1992). If you don't mind the math, you might consult Gregory L. Matloff's *Deep-Space Probes*, 2nd edn (Springer–Praxis, Chichester, UK, 2005).

A few authors have considered what it would take to construct a space–time singularity to allow travel through hyperspace. A very readable presentation of the challenges involved is Adrian Berry's *The Iron Sun* (Warner, New York, 1977).

Possible solar-sail missions to the Sun's gravitational literature have received a great deal of attention in the technical literature. One source is J. Heidmann's and C. Maccone's, "AstroSail and FOCAL: Two Extrasolar System Missions to the Sun's Gravitational Focus," *Acta Astronautica*, vol. 35, pp. 409–410 (1994).

For a recent non-technical review of Project Orion, please see George Dyson's *Project Orion, The Story of the Atomic Spaceship* (Henry Holt & Company, New York, 2002).

7

BREAKING OUT INTO SPACE: VISIONARY FUTURES

I heard the trailing garments of the Night
    Sweep through her marble halls!
I saw her sable skirts all fringed with light
    From the celestial walls!

Henry Wadsworth Longfellow, from *Hymn to the Night*

AH, if we could only follow our instincts. To seek new green lands, free of war, terrorism, overpopulation and pollution—what a wonderful dream!

Conventional wisdom says that the Earth is filled up, and a large self-sufficient lunar colony might be impossible due to lack of water and other essential volatiles. Mars, barely habitable, might be altered to support terrestrial life, but at a Herculean effort and enormous cost over millennial duration. Conventional wisdom is wrong!

One person who balked at this sad notion of reality was Princeton University physics professor Gerard K. O'Neill. Shortly after the successful completion of the Apollo Moon landings, O'Neill presented a class with a challenge: Is a planetary surface the best location for an expanding technological civilization?

Through the 1970s and thereafter, O'Neill and his followers elaborated upon the findings of this small study group. The result of these studies is both surprising and intriguing. For humanity to survive and expand, and for all people to have the opportunity to live in the style of contemporary Americans and Europeans, much of the world's industry might ultimately move into space. And much of the world's population might ultimately relocate to city-states orbiting the Earth or located in the deep night of the extraterrestrial abyss.

Following O'Neill's proposal, humans might begin their expansion into the universe from structures not much larger than the contemporary space station. Utilizing lunar and asteroidal resources, space pioneers would then construct space cities to house thousands (or millions) of terrestrials in approximate Earth-like conditions. As these space habitats are constructed, other teams of space workers would utilize lunar and asteroid resources to construct a series of solar-powered satellites. O'Neill's vision was for these huge stations to beam copious amounts of solar energy back to Earth, thereby supplying an economic base for the in-space population. Sadly, economic studies now show that terrestrial-based power generation may be a more cost-effective solution for quite a while. However, for providing power to a space-based civilization, there is no better source than the Sun. In time, if low-cost Earth-to-orbit transportation modes can be developed, the off-planet human population in O'Neill's colonies could become a significant fraction of the human population back on Earth and be more cost-competitive with ever-dwindling terrestrial energy sources. Even though this has not yet come to pass, O'Neill's concepts stand as the first comprehensive blueprint of humanity's break-out into solar and galactic space.

FIGURE 7.1 An O'Neill Cylindrical Space Habitat. The smaller structures are agricultural cylinders. (Image courtesy http://members.aol.com/oscarcombs/gallery.htm)

CYLINDER CITIES

The most exciting parts of the concept were the detailed plans for comfortable, Earthlike cylinder cities in space. A representation of one of these is presented as Figure 7.1.

Even if these space habitats had all the comforts of home, what would motivate thousands or millions of terrestrials to abandon their home world for greener, albeit artificial pastures? The answer to this question might be found in the social milieu of the 1970s. The disastrous war in Viet Nam was finally winding down; the environmental movement was at its peak. In a masterfully prepared document, a group of scholars dubbed "The Club of Rome" outlined a dire future for humanity in the first quarter of the twenty-first century. Population would be rising at a near-exponential rate, increasing the likelihood of deadly plagues and wars.

As the Third World developed, fossil energy use would increase. This would lead to increased energy costs. Carbon-dioxide emissions from these fossil-fuel plants would result in a "greenhouse effect" which would raise the temperature of Earth's atmosphere. Monster storms would result, which would flood coastal cities and erode shorelines everywhere.

A growing rift would develop between rich and poor nations. Some residents of less-favored regions might actually resort to extreme measures in an attempt to close the gap.

Pessimistically, many environmentalists believed that our wealthy civilization could not withstand the onslaught of this multifaceted "time of troubles." O'Neill offered the political leadership of 1970s Planet Earth a chance to head off disaster. They may be damned by future generations for roundly ignoring him!

Construction of O'Neill's cylinder cities would require the launch into space of several million kilograms of tools and construction gear. Most of the billions of kilograms required for structure, cosmic-radiation shielding, and life support would come from cosmic sources.

There were three basic models for these space cities.

1. Each would consist of a pair of counter-rotating cylinders (to avoid precession).
2. Each would face the Sun and be equipped with adjustable window reflectors so that the Earth day/night cycle could be duplicated.
3. People, animals, and plants on the interior walls of the cylinders would experience one-Earth gravity, produced by the cylinder's rotation.

Model 1 was to be 1 kilometer in length and 200 meters in diameter. Its mass would be 500 million kilograms and it could house 10,000 people. Onboard solar power and agriculture would render the habitat essentially independent of the Earth (at least in regards to energy and life support) shortly after its construction. Models 2 and 3 would successively be larger and 10 times more massive, with larger populations and increasingly effective duplication of terrestrial conditions.

Because the Moon was initially thought to be the ideal resource base for the O'Neill habitats, they were to be stationed at one of the Lagrange points (L4 or L5) in the Earth–Moon system. Leading or following the Moon by 60 degrees, L4 or L5 are gravitationally stable. An object placed in these positions stays there with a minimum of course correction.

Although it is theoretically possible to locate an O'Neill space city anywhere in the solar system, there was another reason for a near-Earth location. This was proximity to the solar-powered satellites that would provide the habitat's economic base.

POWER FOR THE EARTH

As the futurists associated with the "Club of Rome" predicted decades ago, it's crunch time for terrestrial energy reserves. Our appetite for fossil fuel seems unlimited, just at the time when environmental effects are becoming very evident. And, especially in light of the rapid industrialization of populous China and India, these resources are not infinite.

Nuclear fission has its limitations, in part because of nuclear-waste disposal issues, and in part because of fears of terrorism and nuclear-weapon proliferation. Many countries are turning to renewable energy sources—solar and wind—to fill the gap. But these, too, have issues: they are diffuse and intermittent. Nuclear fusion might some day become available, but it always seems to be a few decades in the future.

Advocates of space solar power would employ the space-habitat workforce to address this terrestrial energy crisis. Lunar or asteroid resources would be mined to obtain the construction material for huge power stations located in geosynchronous orbits, always positioned above the same location on Earth's equator.

Try to picture these enormous, albeit very flimsy structures. Thin films of solar panels would be arrayed in kilometer-dimension strips. At the

FIGURE 7.2 A space solar-powered station. (Image courtesy http://members, aol.com/oscarcombs/gallery.htm)

center of the multi-million-kilogram structure would be a microwave transmitting station (Figure 7.2). About 20% of the sunlight striking the array would be converted to microwaves. Because the microwave transmitter would operate at a wavelength for which the atmosphere is transparent (even in cloudy weather), most of the transmitted energy would reach receiving stations (rectennas) on the ground and be distributed to customers through the energy grid.

Each solar station could transmit a billion watts. Appearing to be stationary 36,000 kilometers above Earth's surface, space solar-powered stations would be easily viewed in the night sky. As more of these are built, a linear constellation of bright, stationary "stars" would seem to encircle the Earth.

Constructing the hundreds or thousands of space solar-powered stations required to replace Earth's fossil-fuel and nuclear-fission power plants would be a monumental task requiring decades to complete, and this is where the economic basis for the plan begins to break down—at least in terms of our current ability to forecast such things. As the National Research Council's review of NASA's plan for developing this capability states in its 2001 report titled, "Laying the Foundation for Space Solar Power: An Assessment of NASA's Space Solar Power Investment Strategy",

> The committee has examined the SERT program's technical investment strategy and finds that while the technical and economic challenges of providing space solar power for commercially competitive terrestrial electric power will require breakthrough advances in a number of technologies . . .

Among them is the currently prohibitively high cost of Earth-to-orbit transportation.

Even making the assumption that such transportation is virtually free, the investment required for a mature space-to-Earth power infrastructure will result in an end-user cost that will probably not be competitive with terrestrial sources for quite a while. Alas.

What, then, is the likely future of a human space population? First, the current path for science and exploration will likely continue at a snail's pace for the next quarter of a century at least. Robotic explorers will visit other solar-system worlds with ships of increased complexity and capability, returning reconnaissance that will be useful for the human waves to follow. Human expansion into space will resume after a 30-year hiatus post-Apollo, and this time the missions will include players from multiple nations instead of only the two former competing superpowers. (See Chapter 17 for more information about NASA's current human exploration plans.) As this slow expansion occurs, new technologies will be developed that will enable the cycle of "heavy-lift, high-cost launch" to be broken. Technologies that take advantage of all that nature has to offer will incrementally supplant the more expensive, resource-hungry approaches we currently use. In the next 50 years we will undergo a philosophical and technological transformation from an approach that states "bring it all with you into space" to one that asks "what can we use that is already out there?" Thus will begin the next phase of human expansion and opportunity.

We should not expect to see the capability for a human diaspora within our lifetimes. Barring a technological or economic miracle, the mass migration of humans into space will not happen until, perhaps, our chidren's generation is mid-life. Whether propelled by greed (seeking the riches in the asteroid belt), self-preservation (deflecting the planet killing asteroid before it hits the home world), or political competition (as was Apollo), human economic and political processes will determine the pace and extent of the migration.

BEYOND THE SPACE CYLINDERS

Much interest was sparked by O'Neill's grand scheme. After several technical meetings, a study group was convened at NASA's Ames Space Flight Center in San Francisco and co-convened by Stanford University.

Many of this study's conclusions were in congruence with O'Neill's earlier vision. Construction of large space structures was possible, as was meteorite protection, closed ecosystems and space solar power.

But there was one significant divergence. O'Neill's colonies had walls and internal atmospheres thick enough to shield habitat dwellers from most forms of space radiation. But only the largest 10-kilometer-long O'Neill Model 3 habitats had enough atmosphere to give protection against high atomic-number galactic cosmic rays.

Unless some form of magnetic radiation shielding is developed, the Ames–Stanford Study participants calculated that the masses of the smaller habitats should be increased by at least a factor of 10 to reduce galactic radiation to near-terrestrial levels.

Study participants also considered a number of alternatives to the spinning O'Neill cylinders. One of these was the doughnut-shaped Stanford Torus (Figure 7.3).

FIGURE 7.3 *The Stanford Torus, with solar reflector and radiation shielding. (Image courtesy http://members.aol.com/oscarcombs/gallery.htm)*

LIFE IN THE SPACE CITIES

No matter what the final habitat design, it should not be expected that space dwellers will live under exactly the same conditions as their stay-at-home relatives. Since maximum linear distances between destinations will be no more than a few kilometers, bicycles and electric trolleys will replace automobiles. Because agricultural regions will be very close to population centers, food will always be fresh and refrigeration may be unnecessary.

It is debatable whether food-preparation and service will be communal or up to individuals. Because of concerns regarding fire in the confines of a space city, solar-concentrators or microwave ovens might be the preferred method of food production.

Paper production would be local, owing to the proximity of agricultural regions to centers of industry and population. One import from Earth might be computers—and e-publications might well replace the printed word.

Medicine and dentistry would be significantly different from their terrestrial equivalents. Anesthetics might be generated using products grown in onboard rainforest environments. High-tech equipment would likely be imported from Earth, at least in the initial phases of space settlement and industrialization. If a person suffered from a heart ailment, he or she could move to a lower gravity community closer to the space city's axis of rotation.

Variable gravity within these communities will alter human sports, performance arts, and recreation. Low-gravity dance, gymnastics and hang-gliding have been suggested.

The space habitats will be too small for severe weather patterns to develop. Only in the large Model 3 cylinder cities would clouds form and occasional rain fall from the sky.

O'Neill was impressed by the population-density similarities between space cities and Italian hill towns like Assisi and Siena, with histories dating to the Etruscans or Romans. Although this similarity might promote a great deal of social cohesiveness, it also has a down side. Except when they were united by the Roman Empire or the modern Italian government, many of these idyllic-seeming communities were engaged in constant warfare. It would be unfortunate if rival space city-states warred over claims to particularly rich asteroids or comets.

If one surveys the interior design concepts for the initial space habitats, he or she cannot help but be struck by the wide variety of possibilities. Some designs resemble small university towns like Princeton or Stanford; others are suburban. A few are hippie communes, and even some shopping-mall-like environments have been proposed.

Many of these interior-design possibilities, as well as the matter of space-city governance, should be left to the eventual space-city inhabitants. It is clear, however, that given a non-solar energy source and a method of propulsion, these space habitats will be able to engage in interstellar voyages of centuries or millennia.

FURTHER READING

The dire Club of Rome predictions for our terrestrial future are discussed by J. Forrester in *World Dynamics* (Wright-Allen Press, Cambridge, MA, 1971). Gerard O'Neill's response is discussed in his *High Frontier* (Morrow, New York, 1977).

Many authors have worked off O'Neill's grand scheme. In *The Fertile Stars* (Everest House, New York, 1981), Brian O'Leary presents the case for mining near-Earth asteroids and comets rather than the Moon, to provide a space resource base.

If you are interested in the Stanford Torus, consult *Space Settlements: A Design study*, a 1977 NASA report (NASA SP-413), which was edited by R.D. Johnson.

8

THINKING
INTERSTELLAR

I will find out where she has gone,
And kiss her lips and take her hands;
And walk among long dappled grass,
And pluck till time and times are done,
The silver apples of the Moon,
The golden apples of the Sun.

William Butler Yeats, from *The Song of Wandering Aengus*

THE challenge is daunting, it is perhaps the greatest challenge the human race will ever experience. Sending a spacecraft to a nearby star is not as "simple" as sending one to another of the nine planets in our solar system—a feat, by the way, that we have not yet achieved. (Pluto has not yet been explored, though a mission to do so dubbed, "New Horizons", is on its way to that distant planet.) Why is it so challenging? The simple answer is distance—distances so vast that they cannot be easily comprehended; distances so large that we have to invent new units of measure in order to even readily discuss them.

Before attempting to discuss how humanity might traverse the chasm between the stars, we should discuss the tremendous distances right here in our own solar system. These are the roads we must conquer before we can really make a serious attempt to go to other systems.

EARTHLY AND UNDERSTANDABLE UNITS OF MEASURE

We humans like to measure and then categorize things by size or some other defining characteristic. To do this in a way that promotes understanding and, more importantly, commerce, some sort of standardization is required. In the ancient world, and in the United States today, we measure things using "English Units," though the modern-day English now use the metric system—but more on that later. Not really useful for our discussion of interstellar travel, but relevant because of its history (and arbitrary origins), is the "inch." The inch, as a unit of measure, is based on the width of the average human thumb. The "foot" is based on the length of the average human foot and the "yard" is the distance from the tip of the nose to the tip of the middle finger. A much more useful measure for our purposes is the "mile," though its origin is no less arbitrary: a mile was originally defined to be the distance covered by a Roman legion after taking 1,000 paces.

Modern standards of measure now prevail, and the "mile" is now more precisely defined as being 5,280 feet, the "foot" is 12 inches and the "inch," well, the inch is now defined as being roughly 2.54 centimeters (0.01 of a meter). The centimeter is part of the "metric system," which is actually much easier to use and is based on the distance traveled by light in a vacuum. The 17th General Conference on Weights and Measures in 1983 defined the meter as the distance that makes the speed of light in a vacuum equal to exactly 299,792,458 meters per second.[1] The speed of light in a vacuum is one of the fundamental constants of nature. Since the speed of light defines the meter, experiments made to measure the speed of light are also interpreted as measurements of the meter. The meter is equal to approximately 3.2 feet or 39.3 inches.

In twenty-first century America, most people define relatively nearby distances in units of both distance and time. For example, how many times have you heard people say that "the beach is about a 2-hour drive from

---

[1] Kleppner, D., "On the Matter of the Meter," *Physics Today*, March 2001.

here," or "my aunt lives about 3 hours from here." They are, of course, referring to the time it takes a car, traveling on nearby roads under average traffic conditions and usually obeying the speed limit, to reach a stated destination. For illustrative purposes—one that we will call upon again later—let's say that the distance from City A to City B is 100 miles (62 kilometers). Traveling on an interstate highway with a 70-miles-per-hour speed limit and not taking into account any of the normal driving distractions, this journey will take 1.4 hours. Driving from New York to Los Angeles (2,800 miles), across the same continent that used to take early American pioneers the better part of a year to traverse, the average American can now achieve the distance in 40 hours (assuming no rest, bathroom, food, or gasoline stops!).

On a global scale, the human species seems to have conquered distance. The Earth's circumference is 24,902 miles. If we assume that the Earth has an interstate highway around its equator, then traveling around the world at a speed to which the average person is accustomed (in their car), would take 355 hours. Of course, modern air travel makes such a notion quaint. If one were to take such a journey traveling with an air speed of 500 miles per hour, this around-the-world jaunt would take merely 50 hours. Taking our latest technological leap and assuming you can readily get on board the International Space Station, which is circling the Earth at approximately 17,000 miles per hour, the journey would take merely 1.5 hours. Those poor Roman soldiers, from whom we derived the definition of a "mile" to begin with—assuming they can march at a rate of 15 minutes per mile—would take 259 DAYS (non-stop) to cover this distance.

Does this mean that we humans have conquered "distance" and that extending this capability across the solar system and to the stars is just a simple extrapolation of what we've achieved here on Earth? Sadly, the answer is "no."

DISTANCE IN THE SOLAR SYSTEM AND BEYOND

Unfortunately, we quickly lose our ability to understand, let alone comprehend, distances relevant in describing the scale of our solar system. The Earth is approximately 93,000,000 miles from the Sun. It we could take the Space Station and send it sunward, still traveling at 17,000 miles per hour, it would take 227 days to reach its destination. And sending it on a direct trajectory to the Sun is not possible due to the fact that everything in the solar system, including the Earth, is in orbit around the Sun, thereby making any craft follow a curved path, or trajectory, as it spirals to its destination—which can dramatically increase the total distance to be traversed. Since it is arduous to calculate a trajectory, and hence the actual distance to be traveled, and since this only makes the distance problem worse, we will ignore it for illustrative purposes without compromising the intent of our discussion.

Mars is about 1.5 times as far from the Sun as is the Earth. Since Earth and Mars have different orbital periods (the time it takes to complete one revolution around the Sun), the distance between the two planets varies between approximately 40,000,000 miles and 235,600,000 miles. A spacecraft launched from the Earth to explore Mars takes approximately 9 months to cover this distance on an optimal trajectory. Now we have a realistic interplanetary distance and a realistic timeframe based on actual flight mission history on which to base our discussion. Nine months is a long time, but not so long as to be discouraging. Our continental ancestors took this period of time to get from the East to the West coasts, and they did it readily. Surely we are now on the way to making the solar system our backyard! Not so fast...

Jupiter, the largest of all the planets, is 482,500,000 miles from the Earth, or 3.5 times farther away than Mars (a straight-line distance, not in terms of a required trajectory). The recently completed Galileo mission, following an optimized trajectory, took 5 years to reach its destination! Pluto, the most distant planet in the solar system, is 39.5 times the Earth–Sun distance from the Sun. (A straight-line distance of approximately 3,666,000,000 miles.) The New Horizons mission to Pluto launched in 2006 will not arrive at Pluto until 2015!

It is at this point that most people's eyes begin to fog over and they lose comprehension and interest in understanding the distances involved. After all, who can experientially understand the difference between 3.6 billon miles (the Pluto–Sun distance) and a mere 93 million miles (the Earth–Sun distance)? We then throw into the mix the knowledge that the distance to the nearest star, in the Alpha Centauri system, is 24,790,771,495,138 miles away—and that is the distance to the *nearest* star! To put that in units of millions of miles, it is approximately 24,000 million million miles away. Now we've all lost our understanding of the distance and confusion and despair seem a likely outcome.

Fortunately, there are ways to comprehend this. One way is to use the mathematical construct of logarithms. On a logarithmic scale, each incremental unit of distance is 10 times longer than the previous one. Using this notation, the distance from the Sun to the nearest star system

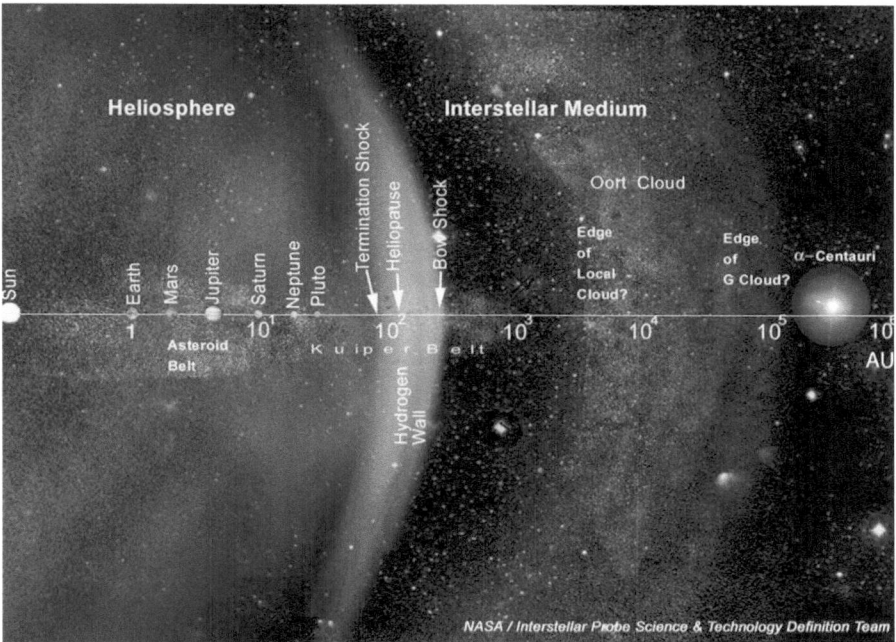

FIGURE 8.1 A logarithmic representation of nearby interstellar space. Each increment of distance represents 10 times more total distance than the previous one.

can be seen fairly easily (Figure 8.1). An alternative method to visualize these immense distances is to create our own version of the solar system, using units of measure that we can readily comprehend. First, we will conveniently "shrink" the Earth–Sun distance to 1 foot. If the Sun is at the center, then the Earth is 1 foot away from it. Mars is conveniently located just 1.5 feet away, Jupiter 5.2 feet, and distant Pluto 39.5 feet. Okay, surely we can now comprehend the distance across the gulf between the stars to our nearest stellar neighbor? On this scale, Proxima Centauri is 50 *miles* away!!! Astronomers long ago were challenged by understanding these distances and created a new unit, called the "Astronomical Unit," to represent the Earth–Sun distance. To convert our "39 foot" solar system into these terms is trivial: replace feet with Astronomical Units, or "AUs," and you have it: the Earth is 1 AU from the Sun, Jupiter 5.2 AU, and Pluto 39.5 AU. Proxima Centauri is then approximately 250,000 AU from the Sun. Now we finally realize that the distance between stars is HUGE. The question becomes: "Can we, the descendants of those Roman soldiers who marched across Europe, cross this vast distance?" In 1999, NASA asked this very question.

NASA's Interstellar Initiative and the Interstellar Probe Mission

In early 1999, NASA's then-Administrator, Dan Goldin, tasked the Office of Space Science and NASA's propulsion experts to explore the possibility of sending a robotic probe to the edge of our solar system and beyond within the next few decades. A series of workshops were held at NASA's Jet Propulsion Laboratory to assess the state-of-the-art in spacecraft propulsion and assess them against the requirements of making a voyage into nearby interstellar space and, if possible, to the nearest star.[2] As discussed elsewhere in this book, there are very few technological options available for true interstellar flight. Though physics tells us that a ship propelled by fusion, antimatter or sails may be able to make the journey, the engineering challenges they pose are far beyond anything humanity can hope to achieve at any time soon. This is NOT the case for missions beyond Pluto into *nearby* interstellar space. In fact, logical extrapolations from today's technology in ion propulsion and solar sails may make such a journey possible within a human lifetime.

To send a spacecraft beyond the edge of the solar system and into interstellar space within the professional lifetimes of those that build and launch it (20–30 years), a suite of instruments on board a small spacecraft will need to travel further and faster than any spacecraft flown previously. A goal of reaching 200 AU within 15 years of launch was established for a mission called, "Interstellar Probe."[3]

A team of experts from NASA examined several relatively mature propulsion options for performing such a mission. The options included an all-chemical rocket approach with planetary gravity assists, solar sails, and electric propulsion. More advanced technologies were not considered due to their relatively low maturity. After all, the goal is to launch something within the next few decades. For a discussion of the relative maturity of various technologies, please refer to Chapter 9.

With the successful flight of the solar-powered ion propulsion system on the Deep Space 1 mission, and the numerous electric propulsion

---

[2] Gavit, S.A., Liewer, P.C., Wallace, R.A., Ayon, J.A. and Frisbee, R.H., "Interstellar Travel—Challenging Propulsion and Power Technologies for the Next 50 Years," *Space Technology and Applications International Forum—2001*, edited by M. El-Genk, AIP Conference Proceedings 552, Albuquerque, N.M., February 2001.

[3] Mewaldt, R.A., Kangas, J., Kerridge, S.J. and Neugebauer, M., "A Small Interstellar Probe to the Heliospheric Boundary and Interstellar Space," *Acta Astronautica*, vol. 35, Suppl. (1995).

systems in use by Earth-orbiting spacecraft, it was now possible to consider the use of electric propulsion for more ambitious deep-space applications.[4] However, solar power is not a feasible option for an electric propulsion system on a fast mission into interstellar space—the spacecraft does not spend enough time close to the Sun to build up adequate velocity—and the farther you get from the Sun, the less power you can generate to drive your propulsion system. The Interstellar Probe study team estimated that a mission to the outer solar system would require a solar collector with a mass estimated to be in the order of 50,000 kilograms in order to collect enough power for an 11-year trip time with electric propulsion. Given the relatively high power levels required to achieve fast trip times, a small fission reactor capable of producing tens of kilowatts of power would be ideal.

Electric thrusters for interstellar precursor missions, regardless of how they receive electrical power, will have to operate at substantially higher specific impulse (approximately 9,000 to 14,000 seconds, depending upon the specific requirements of the mission) to achieve desired trip times. Recall that "specific impulse" is just another way to describe the overall efficiency of the propulsion system. Specific impulse can be defined as: *the amount of time 1 kilogram of propellant lasts in a thruster that is using the fuel to continuously produce a force equal to the weight of 1 kilogram.* The higher the specific impulse, the more efficiently the onboard propellant is converted into thrust. Furthermore, very large amounts of propellant will be required to drive the craft. Ion thrusters currently use xenon, an inert gas, as the propellant of choice. This mission, if undertaken, would require more xenon than is currently produced annually in the world! Not that making xenon is impossible, but either the production capacity would have to dramatically increase, or alternative fuels such as krypton would have to be considered.

In addition to electric propulsion, solar sails seemed to be very promising. Recent developments in lightweight materials and thin films make large-diameter sails for spacecraft propulsion a viable option. The Russian Znamya program resulted in the deployment of spinning reflectors from their Progress vehicles. These reflectors were very similar in size and shape to solar sails, although their intent was to demonstrate an ability to deploy large reflectors capable of illuminating northern Russia during the long winter nights. NASA's Inflatable Antenna Experiment deployed from the Space Shuttle in 1996 demonstrated that large, lightweight structures might

---

[4] Rayman, M.D., Varghese, P., Lehman, D.H. and Livesay, L., "Results from the Deep Space 1 Technology Validation Mission," *50th International Astronautical Congress, Amsterdam, The Netherlands*, 4–8 October, 1999, IAA-99-IAA-11.2.01.

be deployed in space. These are but two examples of recent space demonstrations of solar sail and solar sail component technologies which, with others, are described in more detail in Chapter 13.

The primary performance parameter for solar sails is their areal density (grams per square meter), which determines the acceleration of the sail (i.e. solar pressure [newtons per square kilometer] divided by areal density [grams per square meter] gives acceleration). Areal density is determined by the thickness and density of the sail material, and the mass of the supporting structure. The term "loaded areal density" refers to the entire spacecraft mass, including the payload, divided by the sail area. Solar sail areal density requirements range from around 13 grams per square meter to perform near-term demonstration missions to around 1 gram per square meter for missions beyond the edge of the solar system. For reference, three raisins weigh about 1 gram. Imagine squashing these raisins and stretching them out to cover an area 3 feet by 3 feet and you will have some idea of what the 1 gram per square meter solar sail spacecraft must weigh in order to carry out the Interstellar Probe mission.

The Interstellar Probe study team assumed the development of a spin-stabilized solar sail with a loaded areal density of 1 gram per square meter, and a diameter of approximately 400 meters. The spacecraft would be centered in the 11-meter-diameter central aperture of the sail. A single Delta 2 class launch vehicle would be used to deliver the spacecraft to an Earth-escape trajectory. Once the solar sail is deployed, the spacecraft would spin-up and deploy the sail. The spacecraft would then be placed on a heliocentric (Sun-centered) trajectory from the Earth inbound to a 0.25 AU close solar approach (perihelion). At perihelion, the sail would be oriented to maximize the reflection of sunlight, getting a boost 16 times larger than would be possible at the Earth—thereby sending it on a rapid journey out of the solar system. The increased sunlight pressure at 0.25 AU is due to something called the "inverse square law."

Light emitted from a single source, like the Sun or a campfire, spreads out equally in all directions. As the distance from the source to the sail (or any other object) increases, the amount of light from the source spreads out over a larger and larger area—and the resulting propulsion derived from the sunlight falls accordingly. A more detailed discussion of solar sails and how they operate can be found in Chapter 13.

After reaching 5 AU, the solar thrust available from the sail will become rather low; the team assumed that the sail would be then jettisoned, leaving the spacecraft on a trajectory toward deep space.

Another mission scenario examined by the team used chemical propulsion with planetary gravity-assist maneuvers. Complex trajectories using multiple gravity-assist (GA) maneuvers at Jupiter (J), the Earth (E), Venus (V), and powered solar flybys (S) were considered. The launch vehicles considered ranged from the relatively inexpensive Delta 2 to the costly and high-performing Titan IV with a Centaur upper stage. The team concluded that an all-chemical propulsion approach could achieve the 200 AU mission goal in approximately 25–30 years from launch.

Nuclear-electric propulsion (NEP) was considered to be the only other propulsion option that could both be brought to a sufficiently high technology readiness level and provide fast (10–15 year) trip times in the near term. The evaluation of nuclear-electric propulsion systems for the mission focused on an advanced fission reactor-based concept. It is worth noting that the United States does not currently have a fission reactor that can be used in space.

A nuclear-electric propulsion system would use a nuclear reactor capable of providing up to 200 kilowatts of electricity, launched "cold"— meaning that no radioactive fission byproducts have yet been formed, thereby minimizing the launch risk. The reactor would only be activated to power a Krypton-fueled ion propulsion system when the spacecraft was already on a trajectory away from the Earth. The propulsion system would carry the interstellar spacecraft on its way outward.

In response to the Administrator's challenge, and the results of the various mission and propulsion assessments from the engineering and scientific communities, NASA established the "Interstellar Propulsion Research" Project at NASA's Marshall Space Flight Center. The author (Johnson) was the first and only manager of the Project, which was funded at approximately $2 million per year for the 2 years of its existence. While we did not achieve the goal of developing either an interstellar-capable solar sail or a viable nuclear-electric propulsion system, the project initiated the more robustly funded In-Space Propulsion Technology Program which is maturing solar sail propulsion technology toward its first mission use and the Project Prometheus Program which, for a time, made considerable progress in the development of nuclear-electric propulsion.[5]

*   *   *

How does one get to be manager for NASA's Interstellar Propulsion Technology Project? In my case, I happened to be at the "right" place, at the "right" time, have the "right" technical background, and, most

---

[5] Johnson, L., "Interstellar Propulsion Research Within NASA," *52nd International Astronautical Congress, Toulouse, France*, 1–5 October, 2001, IAA-01-IAA-4.1.02.

importantly, the interest in making it happen. At the time the project came about, NASA, and in particular, the Marshall Space Flight Center, were being resoundingly criticized for not "thinking beyond the space shuttle" with regard to advanced space propulsion technologies. In response to this, the "Advanced Space Transportation Program" (ASTP) was formed and given a modest budget to pursue new technologies and systems that would primarily benefit new Earth-to-space launch systems. A very small part of the budget, about $5 million was allocated annually to propulsion technologies that operated only "in space," meaning that they do all the propulsion after the launcher does its job.

The ASTP had been operating for about a year when Mr Goldin challenged the agency to plan an interstellar mission. He asked NASA JPL to take the lead for the mission and for NASA Marshall to be the lead for the propulsion system. While this was happening, I was Principal Investigator for a small tether propulsion experiment funded by the ASTP called, "ProSEDS." I was therefore in and around the management offices where I would hear snippets of discussion surrounding the interstellar initiative and how Marshall should respond to the Adminis-trator's request. Many of the very good engineers at Marshall openly ridiculed the idea of "interstellar travel," knowing full well the challenges such missions would have to overcome. Being a physicist, and not an engineer, I was perhaps less-biased against the engineering challenges and based my world view more on the underlying physics—which clearly said that such a mission was possible. I was intrigued.

When the first workshop was planned at JPL, I asked if I could attend with other propulsion experts from the Center; and I did. Futuristic, but technically feasible ideas (some day) such as fusion and antimatter propulsion, were discussed at the workshop, which promptly alienated the hard-nosed propulsion engineers from Marshall who were working the very real problems of the space shuttle main engines and in-space chemical upper stages—understandably. To them, it was all a daydream on which they did not have time to waste.

On my return, I gave my impressions to the ASTP manager—an experienced and serious engineer named Gary Lyles. Gary wanted Marshall to support the Administrator's request, but he was very cautious about how to do so. After all, all his other management responsibilities were centered on making very real improvements in relatively near-term propulsion systems and he could not afford to have an "interstellar" requirement give the perception that my other activities were equally "far out."

Now, I should provide some insight as to how I believe my peers in the space propulsion community viewed me at the time. I was an "unknown" physicist without an aerospace engineering degree, whose entrée into advanced propulsion was a system that used a 20-kilometer-long wire, or tether, to extract energy from the Earth's geomagnetic field in order to propel a spacecraft. (To learn more about space tethers, please see Chapter 15.) In essence, I was a proponent of a space propulsion system that would move a spacecraft by pushing on a string. Again, the physics of the idea was sound and even the engineering somewhat known, thanks to the flights of precursor tether missions in the years previous. In other words, I was viewed as being "out of the mainstream" but certainly not a crank!

As I had shown interest in the whole notion of interstellar travel and was viewed as a somewhat "out of the box" thinker, Gary asked me if I would like to take on the challenge as manager for "Interstellar Propulsion Research" within the ASTP. I readily agreed and immediately began working with my science mission counterpart at JPL. It should be noted that my coworkers immediately began, good naturedly, to call me "starman," even making a poster by that name with my picture on it for display in the office area! I had the good fortune of holding this position for about 1.5 years, making the initial investments in solar sails that are coming to fruition today.

*   *   *

The Interstellar Probe mission, as described above, has not yet flown. It is, however, a project on the list of future missions that the space science community wants to see happen. This list of missions, and the (hopeful) date at which they should occur, is called a "science roadmap." This particular roadmap targets the 2020s for launch of humanity's first mission into nearby interstellar space. And yet, the scientists are not satisfied. Recent advances in carbon-based materials, those with inherent thermal properties to allow survival during close solar approach maneuvers, have the potential to dramatically alter our current view of interstellar exploration. Missions to 1,000 AU and beyond can now be considered due to the high velocities achievable by solar sails via close solar approaches. Matloff calculated that a 1-kilometer-diameter sail using a 0.15-AU perihelion maneuver could reach 1,000 AU within 21 years.[6]

If only those Roman soldiers could see what we are trying to do today!

---

[6] Matloff, G. L., *Analysis of Several Candidate Interstellar Propulsion Systems*, NASA Technical Report, MSFC, 1999.

FURTHER READING

The very significant topic of unit definition is treated in many first-year college and university physics texts. One nice treatment is by Hans C. Ohanian, in *Physics*, 2nd edn (Norton, New York, 1989).

A well-researched and very readable account of interstellar-travel research at NASA and elsewhere is Paul Gilster's *Centauri Dreams* (Copernicus, New York, 2004).

The analysis performed by author Matloff on a 1-kilometer-diameter solar sail that could reach 1,000 AU in 21 years was included in an unpublished 1999 Marshall Space flight Center memo to Les Johnson entitled "Analysis of Several Candidate Interstellar Propulsion Systems." Similar calculations have been published in Chapter 4 of Matloff's *Deep-Space Probes*, 2nd edn (Springer–Praxis, Chichester, UK, 2005).

9

TECHNOLOGICAL READINESS

The steel mill sky is alive.
The fire breaks white and zigzag
Shot on a gun-metal gloaming.
Man is a long time coming.
Man will yet win.
Brother may yet line up with brother.

Carl Sandburg, from *The People Yes*

HOW does the manager of a multimillion-dollar space mission know that the technologies required for it are "ready to go"? Often, the technologies required for space exploration are audacious, complex, and very expensive.

- Which are absolutely essential to mission success?
- Which would increase mission performance?
- Are any enabling?
- What about cost?
- How much time and money will it take before the mission hardware can be developed with a reasonable chance of being successful?

These are the questions that must be asked before a decision is made to spend upwards of $250 million on a robotic mission, and much more on a human mission.

In the early space program, most of the technologies required were new. If they were not new *per se* (meaning that some terrestrial application might exist) then the environment in which they would have to operate for a space mission was certainly new—and this environment is what confounds engineers in the development of "space-qualified" hardware. For example, the environment of near-Earth space is a hard vacuum with temperature extremes varying many hundreds of degrees between the times a spacecraft is in sunlight and when it is in shadow. Depending on its altitude, the spacecraft can experience varying exposure from the Earth's Van Allen radiation belts. It can be smashed by a piece of Earth-orbiting debris or a meteroid from outer space—and be converted into a cloud of debris. In addition to the electromagnetic radiation from the Sun (sunlight), the Sun emits penetrating radiation that is especially intense during solar flares—intense enough to "fry" conventional electronics to the point of destruction.

Not only does a technological device have to work as designed, but it has to reliably operate in the most harsh environment humanity has yet encountered—outer space. And this brings us to the question of technological maturity. How does a mission manager know that the required hardware is mature enough to go into space?

In an attempt to provide an "apples to apples" method of comparing the maturity of various technologies, NASA and much of the aerospace community has adopted the "Technology Readiness Level" (TRL) scale, as originally articulated by John Mankins in a 1995 NASA White Paper.[1] The premise is simple, but the application is often arduous and complex, as will be discussed later for a technology known as "electrodynamic tethers."

The NASA TRL scale is intimately keyed to the anticipated operational environment of the space system in question. There are several key terms that must be defined before the rating system can be completely understood:

Relevant Environment—This usually means a simulation of the space environment on the ground, though it might mean an actual space test. Not all systems, subsystems or components need to be tested in full space and launch environments. Others, like solar sails and tethers, need to be

---

[1] Mankins, J.C., "Technology Readiness Levels," NASA Internal White Paper (1995).

tested in space because of their scale and the fact that there is no good way to counter the effects of gravity on the system during a ground test.

Laboratory Environment—This relates to an environment that does not address in any manner the environment to be encountered by the system, subsystem or component during its intended operation. Tests in a laboratory environment are solely for the purpose of demonstrating the underlying principles of technical performance without respect to the impact of environment. Testing at this level can be performed just about anywhere.

Breadboard—This is a hardware test unit that demonstrates function and that does not take into account actual flight-like shape, materials or environment. It has no flight hardware or software and serves only to demonstrate the fundamental operation of whatever is being demonstrated.

Prototype—A prototype demonstrates "form, fit and function." In other words, it looks, acts, and is sized to be just like the space flight version. It is to every possible extent identical to flight hardware and software. It is built to test the manufacturing and testing processes under flight-simulated conditions. The *only difference* between a prototype and a flight unit is that the prototype is meant to be tested realizing that something will need to be changed or fixed as a result of the testing. Very seldom are we smart enough to not learn something during prototype testing that needs to be incorporated into the actual flight unit. After all, that is why we test!

TECHNOLOGY READINESS LEVELS

TRL-1: Basic Principles Observed and Reported

The entry level for technology assessment is Level 1 (TRL-1). To achieve it, one must do little more than demonstrate that a new idea is physically possible, preferably by citing fundamental physical principles and some sort of laboratory experiment as evidence that the fundamental principles are physically real. When an engineer or scientist has an idea and scribbles it on a napkin in the cafeteria over lunch, it has the potential to be TRL-1, but that status will not be granted until the requisite physics is validated and experimental proof provided.

A classic example of a TRL-1 space propulsion system is the proton-fusing interstellar ramjet (see Figure 6.3).[2] All stars, including the Sun, shine by energy generated from the fusion of hydrogen nuclei (protons) deep within the stellar interior to generate heavier nuclei (such as hydrogen) and energy. Stellar thermonuclear fusion has been verified both theoretically and in the laboratory. If (and it's a mighty big if!) only we could learn how to perform this feat in a non-stellar reactor and how to collect sufficient protons from the interstellar medium, then we could advance the TRL of the ramjet and perhaps develop the capability to perform relativistic space voyages.

TRL-2: Basic Principles Observed and Reported

Taking the idea from the white board or university physics colloquium to an engineering concept is essential for maturing to TRL-2. It is at this stage that a notional method of producing propulsion must be shown— one that is based on "real" physics but also on solid engineering principles. It is here that a great many new ideas falter. The physics may say there's a potential application, but unless someone can visualize a way to engineer a laboratory demonstration of the principle, the idea will languish. This level does not require a laboratory demonstration, but such a demonstration must be at least able to be designed—perhaps awaiting a method of implementation.

An example of a space propulsion technology in this category is the antimatter drive.[3] When a chunk of antimatter meets an equivalent mass of normal matter, the result is the total conversion of matter to energy. Matter/antimatter annihilation therefore releases more energy per unit mass than any other known reaction in particle physics and has tremendous energy density. When relatively heavy particles such as protons and antiprotons annihilate, the initial reaction products are a spray of electrically charged, short-lived particles that decay into neutrinos and low-energy gamma rays. The neutrinos pass through normal matter without reacting and the gamma rays will escape the spacecraft's engine immediately unless absorbed by shielding. The charged-particle debris from the proton/antiproton annihilation can conceptually be focused by electromagnetic fields and expelled through the engine's nozzle. The reaction to this charged-particle exhaust produces sufficient thrust to propel a spacecraft to the stars.

---

[2] Cassenti, B., "The Interstellar Ramjet," *40th AIAA/ASME/SAE/ASEE Joint Propulsion Conference*, Fort Lauderdale, FL, July 11–14, 2004.
[3] Forward, R.L., "Antimatter Propulsion," *JBIS*, vol. 35 (1982).

In many terrestrial laboratories, physicists studying the fundamental building blocks of matter routinely produce proton/antiproton annihilation events. Antiprotons, produced at great cost, can be stored in electromagnetic 'bottles" for weeks or months. Although the economics are currently prohibitive, it certainly seems physically possible to construct a demonstration antimatter-annihilation rocket engine that would produce measurable thrust. Please refer to Chapter 20 for a more complete discussion of antimatter propulsion.

TRL-3: Analytical and Experimental Critical Function and/or Characteristic Proof of Concept

It is at this stage that a laboratory demonstration of propulsion must have been demonstrated. In addition, studies must have been done to show that the laboratory-scale experiments can evolve to implementation as a spacecraft propulsion system. It is not sufficient to demonstrate that a new technology can produce thrust. The process by which it produces thrust must be shown to be scalable for real-world applications in spacecraft propulsion. The exact engineering processes for the scalability need not be known, but gross scaling and assessment must be achieved by expert consensus. Being recognized as achieving TRL-3 may be somewhat subjective, as no full-scale systems have yet been designed or demonstrated.

Beamed-energy propulsion (BEP) is arguably TRL-3.[4] In a BEP system, the heavy parts of a rocket (propellant, energy source, even the engine) are left on the ground or in orbit, while the payload and its support structure carry out the mission. Instead of burning onboard fuel to provide the energy to convert propellant into thrust, energy is sent to the spacecraft via a laser, microwave, or particle beam. The energy beam excites or heats the propellant, the expulsion of which produces thrust. Alternatively, the momentum of the energy beam (whether it is light or particles of matter) is exchanged via reflection with the spacecraft to produce thrust.

BEP has been demonstrated in the laboratory, where the interaction distance (the distance across which the energy is beamed) is small and the proximity to the power source is maximized. When one merely has to plug into the national power grid for virtually unlimited power, generating measurable BEP-produced thrust is relatively easy. Some demonstrations have been performed out of the laboratory, in which very large lasers developed for defense applications have boosted kilogram-mass payloads many feet into the air.

---

[4] Sercel, J.C. and Frisbee, R.H., "Beamed Energy for Spacecraft Propulsion," *Space Manufacturing 6—Nonterrestrial Resources, Biosciences, and Space Engineering, Proceedings of the 8th Princeton/AIAA/SSI Conference*, Princeton, NJ, 1987.

There are few known limits to BEP. The engineering challenges, however, are daunting. Megawatts of continuous power for the long periods of time required for deep-space exploration cannot yet be engineered. Notional concepts to produce this infrastructure have price tags of billions of dollars and would produce technologies that could also serve as space weapons. This is not to say that such propulsion concepts are without merit—beamed energy might well be the way humanity reaches the stars in the distant future (see Chapter 21).

TRL-4: Component and/or Breadboard Validation in a Laboratory Environment

It is at TRL-4 that a technologist must begin to think about how a propulsion-producing technology might be developed as a system. A propulsion system is a very different thing from a thruster. The system consists of the thrust-producing element along with all of the other elements that are necessary for the thruster to work. If any single component of the system has not been demonstrated to achieve a given TRL level, then the entire system will remain at the lowest demonstrated component's maturity level.

Low-power (< 1 kilowatt) Hall-effect thrusters are a good example of a TRL-4 technology.[5] A Hall thruster is a type of electric propulsion in which the propellant is accelerated by an electric field in a plasma discharge using a radial magnetic field. For a more detailed description, please refer to Chapter 11. Hall thrusters were originally conceived in the United States and successfully matured for Earth-orbiting spacecraft applications by the former Soviet Union—using high power. Lower power thrusters are of interest for robotic exploration beyond Earth orbit and have yet to be demonstrated with the required operating lifetimes.

A Hall propulsion system comprises the thruster, propellant, propellant feed systems and tankage, power conditioning (modifying the power required to operate the thruster from the solar energy source), and whatever mechanical interfaces are needed to physically bolt the propulsion system to the spacecraft. While engineers have a good idea of the technology readiness for most of these components, there is considerable uncertainty as to whether the Hall thruster itself can operate for very long periods of time—up to many years. Since the overall long-life (yet to be demonstrated), low-power "system" has not yet been assembled and tested

---

[5] Oleson, S.R. and Sankovic, J.M., *Advanced Hall Electric Propulsion for Future In-Space Transportation*, NASA/TM-2001-210676, April 2001.

in the laboratory, let alone a simulated space environment, low-power Hall thrusters are currently considered to be at TRL-4.

TRL-5: Component and/or Breadboard Validation in a Relevant Environment

The next step in the maturation of a propulsion technology is its demonstration in a simulated space environment. As was discussed in the opening paragraphs of this chapter, the space environment is a challenging one. Devices that work perfectly well in a laboratory often fail to function when in space. In order to verify that a propulsion system will function in space, the space environment must often be simulated on the ground. Some parts of the space environment cannot be adequately simulated here on Earth—the effects of gravity is a prime example—but others can be: vacuum, extremes of heat and cold, radiation, and atomic oxygen are but a few. Demonstration in a simulated environment, one that takes into account the salient effects of space on the particular system being tested, is required to achieve this level of technological maturity.

Solar Sail propulsion systems are generally considered to have achieved at least TRL-5.[6] NASA has tested two competing technical approaches to solar sail fabrication, deployment and test in the largest vacuum chamber in the United States. The facility, called, the "Plumbrook Space Power Facility," is located in Sandusky, Ohio, and is part of NASA's Glenn Research Center. Solar sails achieve propulsion by the exchange of momentum from sunlight with a large, thin, reflective and very lightweight sail material physically attached to a spacecraft. Whenever and wherever sunlight is plentiful, a solar sail will be able to thrust. In this case, the "relevant" environment criteria was met for the solar sail propulsion system piecewise: The deployer and sail were exercised under space-realistic thermal vacuum conditions at Plumbrook, with the temperature varying as would be experienced in space, while simultaneously operating in a hard vacuum. The materials used to fabricate the deployer and masts have been previously demonstrated in space. The sail material, however, is new and is currently under a long-duration life test in which it will experience solar ultraviolet illumination concurrently with a realistic simulation of solar particle radiation. This "life test" will be completed approximately 3 years after its start. In addition, to confirm that

---

[6] Lichodziejewski, D., West, J., Slade, K. and Belvin, K., "Development and Ground Testing of a Compactly Stowed Inflatably Deployed Solar Sail," *45th AIAA/ASME/ASCE/AHS/ASC Structures, Structural Dynamics and Materials Conference*, Palm Springs, California, 2004.

the sail will make the transition from atmospheric pressure to vacuum, the full-size sails have undergone an ascent vent test to simulate the pressure change experienced by the system during launch from the ground into space.

Solar sails will remain at TRL-5 until they are demonstrated in space, as it is impossible to fully integrate and demonstrate full-size (60- to 100-meter-diameter) solar sails on the ground in more than a piecewise fashion. It is impossible to eliminate the effects of gravity, which imposes much more severe stress on a fully deployed sail than will ever be experienced in space. In addition, the are no vacuum chambers capable of accommodating sails larger than 20 meters in diameter, thus forever limiting our ability to assess deployment and deployed dynamic characteristics on the ground. Chapter 13 provides a more detailed discussion of solar sails.

TRL-6: System or Subsystem Model or Prototype Demonstration in a Relevant Environment (Ground or Space)

TRL-6 is the usual end point for a technology development program. This is the point at which a flight-like system is fabricated and tested in as close to a space-operational environment as is possible. Some propulsion systems may be able to attain TRL-6 on the ground, but others will require demonstration in space. Mission managers will accept some technologies demonstrated to be at TRL-6 for implementation in a science or exploration mission because, by definition, they have been demonstrated in a "relevant" environment.

The Hall-effect thruster system mentioned in the discussion of TRL-4 above, will eventually be able to be demonstrated to TRL-6 on the ground. There appear to be no limiting physical factors that would prohibit the fabrication of a low-power Hall propulsion system, including the thruster, power-processing, propellant and associated feed system hardware, and testing it under space-realistic conditions on the ground. Large vacuum tanks, much smaller than that at Plumbrook, but large enough to accommodate the thruster plume generated by the Hall system, are plentiful and can be made to simulate the harsh conditions of space. In this case, the effects of gravity on the system's performance are negligible and can easily be accounted for in the testing.

A solar sail propulsion system, however, will require space demonstration to achieve this level of readiness. Only in space itself can the gossamer sail be fully deployed and characterized without the effects of gravity dominating the situation and likely distorting the results. Only in space can a 60- to 100-meter-diameter sail be deployed in a vacuum and experience the extreme thermal conditions inherent with space operation.

It is at this point that it might be interesting to examine a real-world example of a technology maturation effort that was to have demonstrated a new propulsion system to TRL-6. The Propulsive Small Expendable Deployer System (ProSEDS) tether experiment was to have demonstrated in space the use of an electrodynamic tether to produce propulsion from its interaction with the Earth's magnetic field and ionosphere.[7] (For further information on space tethers, see Chapter 15.) In orbit, ProSEDS would have deployed from a Delta-2 rocket's second stage a 3.1-mile-long (5-kilometer), ultra-thin bare-wire tether connected with a 6.2-mile-long (10-kilometer) non-conducting tether. The interaction of the bare-wire tether with the Earth's magnetic field and the ionosphere was to produce thrust, thus lowering the altitude of the stage. (Due to unfortunate circumstances surrounding ProSEDS planned launch date and the tragic loss of the Space Shuttle Columbia shortly before that date, the experiment was canceled.) The story of the ProSEDS tether begs to be told as an example of a TRL maturation "lessons learned."

When it was originally conceived, the ProSEDS tether experiment was to build on the successful heritage of the three missions flown previously, collectively known as the Small Expendable Deployer System (SEDS) missions. All three of the SEDS missions (SEDS-1, SEDS-2 and the Plasma Motor Generator) used essentially the same tether deployer, some with significantly different tethers. The deployer, therefore, was considered to have achieved TRL-6, perhaps even TRL-7. The ProSEDS would use the SEDS deployer to reel out an aluminum wire instead of the SEDS-proven Spectra 2000 polymeric tether. Immediately, the issues began to surface …

The exit guides of the deployer were not sufficiently wear-resistant to sustain the friction of a wire crossing them versus the previously used Spectra polymer and had to be replaced. The core, around which the tether was wound, and associated deployment hardware, were not designed to accommodate a wire with a limited bend radius like that needed for ProSEDS. (Wires will not coil as tightly as a fabric; Spectra has the properties of a fabric.) One has only to pick up a piece of fishing line and compare it to speaker wire to get an idea of the differences between winding and deploying Spectra (which is used in some fishing lines) and

---

[7] Johnson, L., Estes, R.D., Lorenzini, E., Martinez-Sanchez, M. and Sanmartin, J., "Propulsive Small Expendable Deployer System Experiment," *Journal of Spacecraft and Rockets*, vol. 37, no. 2, March–April 2000.

the ProSEDS wire tether. As the team was working to accommodate these changes, the tether team was encountering its own problems.

A bare aluminum wire in orbit around the Earth will get hot. Not only from ohmic heating due to electrical current running through it, but also due to the direct heating from sunlight and, indirectly, from the sunlight reflecting off the earth—known as the Earth albedo. Once it begins to run hot, its electrical conductivity, or efficiency, drops and it also begins to lose its structural strength. To mitigate the strength loss, it was decided to braid the aluminum wire around a Kevlar core. To solve the heating problem, a coating had to be developed to efficiently reflect sunlight while still remaining electrically conductive, which is the reason aluminum was chosen in the first place.

In addition, there was the atomic oxygen problem. When aluminum is exposed to oxygen, a thin layer of aluminum oxide forms on its surface. Aluminum oxide is not as conductive as aluminum, and thereby reduces one of the key performance parameters of the tether. The coating being developed to keep the aluminum tether from running "hot," must be electrically conductive and not susceptible to atomic oxygen. It also has to function when the tether enters the Earth's shadow and the temperature plummets.

It was at this time that the team realized that the TRL of the tether propulsion system was significantly lower than believed when the project began. This was not due to deliberate overstatement of the team's readiness, but rather to an incomplete understanding of the "relevant environment" in which the system would have to operate and how it differed from that which had been flown, successfully, before. But the problems were not over ...

Once a tether that met all the requirements was in-hand, it had to be tested in the newly redesigned deployer. After a series of vacuum deployment tests, it was noticed that the tether coating, when abrasively running across the exit guides of the SEDS deployer, flaked off and left a residue in the deployer. In space, the residual dust would be contaminant and a potential cause of multiple mission-ending events. The team had to go back to the drawing board to develop an electrically conductive, strong, atomic-oxygen-resistant, reflective, and not-too-stiff tether that also didn't leave a residue in the deployer as it was being reeled out. The team did it, but it took time and money that was not originally in the mission budget.

The bottom line messages to the technology community from the ProSEDS experience are these:

1. Define the operational environments early.
2. Don't believe that any hardware can simply be "adapted" from one use to another.

3. Perform a very thorough assessment of each and every bit of new technology being used in the mission.

TRL-7: System Prototype Demonstration in a Space Environment

Had the ProSEDS experiment flown and been successful, and had a follow-on mission application required a tether system with performance or design directly scalable to that system, then ProSEDS electrodynamic tether propulsion system would have demonstrated TRL-7. TRL-7 acknowledges that it is often impossible to fly the same system that is demonstrated in space—hence the word, "prototype." What is flown in space must at least be scalable by modeling or test to the system needed for another mission.

Perhaps a better example of a new, advanced propulsion system that has reached TRL-7 is ion propulsion. Specifically, the NSTAR (NASA Solar electric propulsion Technology Application Readiness) ion engine, which first flew in space about a decade ago, is considered to be a TRL-7 propulsion technology.[8] Developed by NASA and flown on board the Deep Space 1 mission in 1998, the NSTAR ion engine provides highly efficient, low-thrust propulsion ideal for deep space exploration in that it dramatically lowers the propulsion system mass below that which would be required if traditional chemical propulsion were used. An ion engine relies on electrically charged atoms, or ions, to generate thrust. Xenon, an inert, non-combustible gas, is electrically charged and the ions are accelerated to a speed of about 62,900 miles per hour (30 kilometers per second). The ions are then emitted as exhaust from the thruster, creating a force that propels the spacecraft in the opposite direction.

The next mission to use the NSTAR engine will be Dawn, which is to launch by 2007. The engine is not a carbon copy of the one flown by Deep Space 1 and, therefore, is not quite as high on the TRL scale as originally believed. Fortunately, the team working on Dawn is the same team that flew on Deep Space 1 and the effects of the drop in overall TRL were addressed by the mission team—Dawn will use more than one NSTAR engine, requiring changes to the overall power management and distribution scheme, propellant and tankage, and system operations plans. For missions that, in the future, may require more than one NSTAR engine to operate simultaneously (Dawn will fire them serially), multiple

---

[8] Rayman, M.D., Varghese, P., Lehman, D.H. and Livesay, L., "Results from the Deep Space 1 Technology Validation Mission," *50th International Astronautical Congress*, Amsterdam, The Netherlands, 4–8 October, 1999. IAA-99-IAA-11.2.01.

thruster tests will be essential as part of the pre-flight ground test program. The interaction effects between such simultaneously operating thrusters are relatively unknown.

TRL-8: Actual System Completed and "Flight Qualified" Through Test and Demonstration (Ground or Space)

A TRL-8 propulsion system is one that has been shown to perform all mission-level requirements, at a full system level, either on the ground or in space. The flight mission would use the system without modification, other than integrating it within the spacecraft. Most, of the space propulsion systems at TRL-8 are chemical, with a few notable exceptions. Having been the workhorse of space exploration for the past 40 years, chemical propulsion has been repeatedly used successfully on missions in Earth orbit and beyond. A non-chemical system to attain this level of maturity is the arcjet thruster.

Arcjets operate by heating a propellant gas by passing it through an electric-arc discharge instead of heating it via chemical combustion. The amount of onboard propellant used to attain a certain amount of thrust is less with arcjets than with a conventional chemical propulsion system, thereby allowing more payload or longer mission life (by using the mass savings to carry even more propellant). Arcjet thrusters are in commercial use by the satellite communications industry.

CONCLUSION

There you have it. A systematic way of looking at the relative maturity of a technology is in common use by the industry and NASA—and will hopefully let the reader better understand that we now have a quick and easy way to distinguish between relatively immature, but very promising technologies such as those that use "beamed energy" (TRL-3) and electric arcjets (TRL-8).

FURTHER READING

For a not-too-technical introduction to the interstellar ramjet concept and its derivatives, consult E.F. Mallove and G.L. Matloff, *The Starflight Handbook* (Wiley, New York, 1989). A more recent, and more technical, ramjet review is in G.L. Matloff, *Deep-Space Probes* (Springer–Praxis, Chichester, UK, 2000).

The above references also describe the early history and physics of beamed–propulsion concepts. For more recent work on the experimental levitation of small payloads by microwave (maser) and laser beams, consult J. Benford and J. Benford, "Flight of Microwave-Driven Sails: Experiments and Applications," and L.N. Myrabo, "Brief History of the Lightcraft Technology Demonstrator (LTD) Project," which are published in *Beamed Energy Propulsion, First International Conference on Beamed Energy Propulsion, Huntsville, AL, 2002*, ed. A.V. Pakhomov, *AIP Conference Proceedings*, vol. 664 (American Institute of Physics, Melville, New York, 2003).

Aerocapture: A braking maneuver using atmospheric drag to slow an interplanetary spacecraft and capture it into final orbit - using the natural environment of the destination instead of an on-board propulsion system.

Atmospheric drag slows the spacecraft, which must be protected from the atmospheric entry environment. Types of Aerocapture systems include:

- *Rigid aeroshell*
- *Inflatable aeroshell*
- *Thin-film ballutes—combination balloon & parachute*

Aerocapture can:

- *Reduce fuel requirements by 20-80%*
- *Achieve orbit much faster than aerobraking*
- *Reduce trip time from Earth*

Potential destinations enabled or enhanced by Aerocapture include:

- *Mars*
- *Saturn's moon, Titan*
- *Venus*
- *Earth*
- *Jupiter*
- *Saturn*
- *Uranus*
- *Neptune*

Key disciplines for Aerocapture:

- *Aerothermodynamics*
- *Atmospheric modeling*
- *Guidance, Navigation & Control*
- *Trajectory design & performance*
- *Aeroshell structures & materials*
- *Thermal Protection System materials/models*
- *Instrumentation*
- *Systems Engineering & Integration*

Materials technology requirements:

- *High temperature operation*
- *Low mass, high tensile strength*
- *High ultraviolet reflectivity & visable/infrared emissivity*
- *Resistance to reactive planetary atmospheres*
- *Appropriate bonding & insulation materials*
- *Capable of compact stowage & deployment in space*
- *Resistance to space environment effects*

Inflatable Aeroshell
Lockheed Martin

Mid L/D Aeroshell
NASA

Neptune Aerocapture Vehicle
NASA

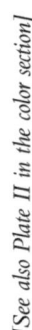

Neptune   Titan

NW-2004-06-060-MSFC

Neptune and Titan aerocapture studies performed by a One NASA team including the Langley Research Center, Ames Research Center, the Jet Propulsion Laboratory, Johnson Space Center, and Marshall Space Flight Center. Funding provided by the NASA In-Space Propulsion Technology Program. www.InspacePropulsion.com

Aeroassist Flight Experiment (AFE)
NASA

Blunt Body Aeroshell
NASA

Hypercone
Vertigo, Inc

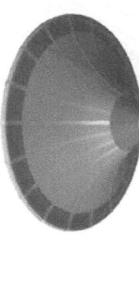

Clamped Ballute
Ball Aerospace

Trailing Ballute
Ball Aerospace

[See also Plate II in the color section]

10

SPACE BRAKES
("LIVING OFF THE LAND" BY USING A PLANETARY ATMOSPHERE)

The upper air burst into life!
And a hundred fire-flags sheen,
To and fro they were hurried about!
And to and fro, and in and out,
The wan stars danced between.

Samuel Taylor Coleridge, from *The Rime of the Ancient Mariner*

FUTURE space explorers should consider a planet's atmosphere as a natural resource to be exploited to allow more affordable and capable space exploration. Planetary atmospheres provide a tremendous resource for explorers who want to minimize the amount of propulsion they must carry from the home planet and, potentially, reduce interplanetary trip times. By allowing a spacecraft to have a controlled entry into the upper atmosphere of a planet, the resulting friction with that atmosphere may replace costly and heavy fuel that would otherwise be used to slow the spacecraft as a means to decelerate and enter into orbit.

There are three ways the atmosphere of a planet may be used to offload propellant and provide a new capability for orbit capture or descent to the surface: aeroentry, aeroassist, or aerocapture. A fourth, aerogravity assist, might use the planet's atmosphere to increase the performance of the gravity–assist maneuver described in Chapter 4. Each has its own challenges, but they all share a common physics and may use very similar technologies. Not discussed here is the notion that the gases within the atmosphere might be extracted and used as fuel for a conventional rocket engine.

AEROENTRY

When a spacecraft enters a planet's atmosphere from space, the friction resulting from the passage of the spacecraft through the relatively thick atmosphere creates friction, converting the kinetic energy of the space-craft's motion (which is typically many kilometers per second, provided to the spacecraft by some other propulsion system or rocket) into heat energy, thus slowing the spacecraft.[1] The space shuttle performs aeroentry every time it returns from orbit around the Earth, as did the Apollo capsules that sent men to the Moon. The space shuttle, which orbits the Earth approximately every 90 minutes as it travels through space at more than 17,000 miles per hour, enters the Earth's atmosphere and uses it to slow down to zero miles per hour as it lands.

The primary technology required for aeroentry is the heat shield. Without some sort of shield, the enormous heat generated during atmospheric entry would quickly melt the spacecraft, or at least critical parts of it. To provide this protection, families of Thermal Protection Systems (TPS) were developed and have been used successfully on multiple vehicles, including the space shuttle as well at the Mercury, Gemini, and Apollo spacecraft.

The maneuver is not without risk, as was tragically seen in the destruction of space shuttle Columbia in 2003. Without its high-performance heat-resistant materials protecting the lightweight aluminum skin of the orbiter, the hot gasses created during its passage through the

---

[1] Regan, F.J., *Reentry Vehicle Aerodynamics* AIAA Education Series, 1984.

atmosphere acted like a blowtorch, damaging the vehicle and ultimately causing it to come apart high above the Earth.

Aeroentry has also been used at Mars. The Viking missions used aeroentry on their way to the surface of Mars in 1976, as did the Mars Pathfinder mission in 1997 and the Mars Exploration Rovers in 2004.[2] In these missions, aerodynamic forces alone were not sufficient to land the spacecraft safely on the surface of Mars. Other systems, like a rocket engine in the last moments of the descent (Viking) or air bags (Mars Pathfinder and Mars Exploration Rovers) were also used. However, if they had not been able to use the Martian atmosphere as a brake, and had been required to perform the descent using only a rocket, the total spacecraft mass at launch might have made the missions impossible to carry out. Simply too much rocket propellant would have been required to accelerate the ship on its journey to Mars and slow it down for landing. Unfortunately, some airless destinations, such as the Moon, provide us with only one option—a rocket-based landing.

In our exploration of the solar system, it will be possible to use aeroentry at any destination that has an atmosphere. This includes Earth, for use when returning from some deep-space destination, Mars, Venus, Jupiter, Saturn, Uranus, Neptune, and Titan—and any other moon blessed with this natural resource.

AEROBRAKING

If a spacecraft is not to land on the surface of a planet, but is rather required to go into orbit around it, then aerobraking can be of significant benefit. Aerobraking significantly reduces the amount of onboard propulsion needed to slow down and be captured into orbit, but it does not eliminate it. To understand why this is so important, it is useful to discuss how an orbiter is captured around a planet using only rocket-based propulsion.

A typical mission *without* the use of aerocapture would go something like this: (1) *launch* (using some sort of rocket to go from the surface of the Earth into space); (2) *Earth escape* (using a propulsion system on board the spacecraft to provide enough velocity to escape the gravitation attraction of the Earth); (3) *braking or orbital insertion* (using the same or a different

---

[2] Spencer, D.A., Blanchard, R.C., Braun, R.D., Kallemeyn, P.H. and Thurman, S.W., "Mars Pathfinder Entry, Descent, and Landing Reconstruction," *Journal of Spacecraft and Rockets*, vol. 36, no. 3, May–June 1999.

onboard propulsion system to shed energy and slow down the spacecraft so that it can be captured into orbit by the target planet's gravitational field; (4) *circularization* (meaning that that spacecraft uses its onboard propulsion system to place it in a useful orbit, typically a circular one).

The difference between this scenario and one that makes use of the atmosphere for aerobraking occurs in the very last phase. Instead of using an onboard chemical propulsion system to modify its final orbit, the spacecraft would use the friction between its solar arrays and the planet's atmosphere to slow down in small increments—a little every orbit until the final desired orbit is achieved. A spacecraft is typically captured into some sort of elliptical orbit and then performs a series of propulsive maneuvers to circularize it. These maneuvers are performed at periapsis (the portion of the orbit where the spacecraft is closest to the surface) because any thrusting there will counter-intuitively lower the spacecraft's apoapsis—or high point of the orbit. With aerobraking, the spacecraft encounters the planetary atmosphere during periapsis, using the slight friction between its solar arrays and the atmosphere to slowly bring down the ellipse toward a more circular orbit. Instead of using propellant, repeated encounters with the atmosphere can perform the same function. The Mars Global Surveyor mission—while reducing the need for significantly more propellant, used this approach in lieu of a rocket.

The primary benefit of aerobraking is the reduction in the amount of fuel needed. Without it, some missions would require larger and more expensive rockets simply to launch the spacecraft with all of the additional propellant on board. A drawback is that several orbital passes are required before the spacecraft can be where it needs to be—sometimes taking months to accomplish! This means that the spacecraft will not be able to achieve its objectives as quickly and it also introduces additional mission risk. Each time an additional maneuver is added, the risk of something going wrong increases.

Aerobraking can take several months to perform, as was the case of the Mars Global Surveyor, which launched in November 1996.[3] The spacecraft's mission was to analyze and send data back to Earth about the planet's magnetic field, atmosphere, and surface. The Mars Global Surveyor made a series of aerobraking maneuvers over a nine-month period to gradually reduce its altitude and achieve its intended orbit. The Mars Odyssey spacecraft, launched in 2001, made a series of aerobraking

---

[3] Beerer, J., Brooks, R., Esposito, P., Lyons, D., Sidney, W., Curtis, H.L. and Willcockson, W., "Aerobraking at Mars: The MGS Mission," *Journal of Spacecraft and Rockets*, vol. 33, January 1996.

Color Section

The Art of Aerocapture

How atmospheric drag can be used to slow an interplanetary spacecraft and capture it into orbit

National Aeronautics and Space Administration

PLATE 1

Aerocapture: A braking maneuver using atmospheric drag to slow an interplanetary spacecraft and capture it into final orbit - using the natural environment of the destination instead of an on-board propulsion system.

Atmospheric drag slows the spacecraft, which must be protected from the atmospheric entry environment. Types of Aerocapture systems include:

- *Rigid aeroshell*
- *Inflatable aeroshell*
- *Thin-film ballutes—combination balloon & parachute*

Aerocapture can:

- *Reduce fuel requirements by 20-80%*
- *Achieve orbit much faster than aerobraking*
- *Reduce trip time from Earth*

Potential destinations enabled or enhanced by Aerocapture include:

- *Mars*
- *Saturn's moon, Titan*
- *Venus*
- *Earth*
- *Jupiter*
- *Saturn*
- *Uranus*
- *Neptune*

Key disciplines for Aerocapture:

- *Aerothermodynamics*
- *Atmospheric modeling*
- *Guidance, Navigation & Control*
- *Trajectory design & performance*
- *Aeroshell structures & materials*
- *Thermal Protection System materials/models*
- *Instrumentation*
- *Systems Engineering & Integration*

Materials technology requirements:

- *High temperature operation*
- *Low mass, high tensile strength*
- *High ultraviolet reflectivity & visable/infrared emissivity*
- *Resistance to reactive planetary atmospheres*
- *Appropriate bonding & insulation materials*
- *Capable of compact stowage & deployment in space*
- *Resistance to space environment effects*

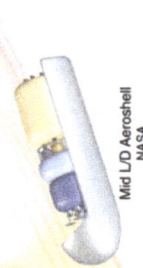

Inflatable Aeroshell
Lockheed Martin

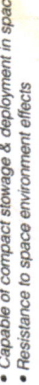

Mid L/D Aeroshell
NASA

Neptune Aerocapture Vehicle
NASA

Aeroassist Flight Experiment (AFE)
NASA

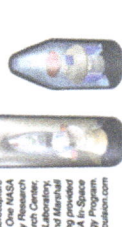

Neptune

Titan

*Neptune and Titan aerocapture
studies performed by a One NASA
team including the Langley Research
Center, Ames Research Center,
the Jet Propulsion Laboratory,
Johnson Space Center, and Marshall
Space Flight Center. Funding provided
by the NASA In-Space
Propulsion Technology Program.
www.InspacePropulsion.com*

NIW-2004-08-080 MSFC

Hypercone
Vertigo, Inc

Blunt Body Aeroshell
NASA

Clamped Ballute
Ball Aerospace

Trailing Ballute
Ball Aerospace

PLATE: II

PLATE III

PLATE IV

PLATE V

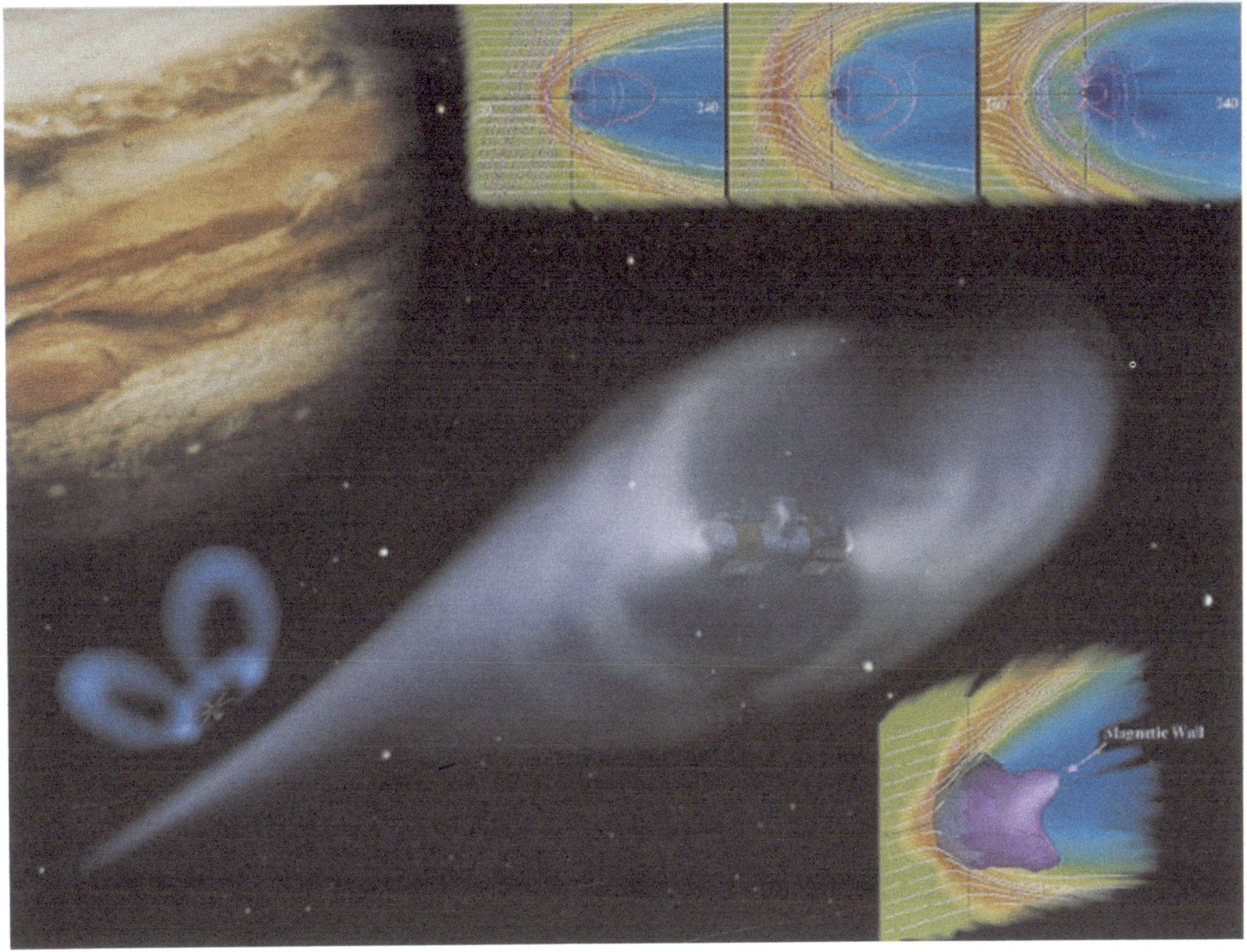

Ischemic Wall

PLATE VI

Advanced Chemical Propulsion

Today's chemical rockets are almost at the theoretical limits of their performance. Researchers are seeking to extend these limits by improving storage life, developing lightweight propellant management systems, and increasing the capability of cryogenic propellant systems to support deep space exploration; and by developing new propellants.

Aerocapture

Aerocapture describes a method of using a planet's atmosphere to assist in decelerating an incoming spacecraft in order to achieve orbit, without the need for on-board propellant. Friction between the spacecraft and the planetary atmosphere slows the spacecraft, allowing it to enter orbit.

Electric Propulsion

Electric propulsion systems use electromagnetic, electrostatic, or electrothermal energy, instead of chemical energy, to accelerate propellant and achieve low, but more efficient, thrust for longer periods of time. The primary benefits of electric propulsion are decreased launch mass, increased delivered mass, and reduced mission time as compared to chemical systems.

Solar Sail Propulsion

Thin, lightweight reflective solar sails produce thrust by harnessing the "pressure" of sunlight. Because they use no propellant, solar sails can open new regions of the solar system for exploration and long-duration observation.

Emerging Propulsion Technologies

The In-Space Propulsion Technology Program actively identifies and evaluates innovative candidate technologies for development. One such technology is the Momentum-eXchange/Electrodynamic-Reboost (MXER) tether, which may provide a reusable in-space infrastructure for high-thrust and high-efficiency spacecraft propulsion. A MXER tether facility would be capable of boosting spacecraft from Low Earth Orbit to higher energy orbits, and could dramatically enhance the performance of all other space transportation technologies while reducing launch costs.

PLATE VII

maneuvers over a period of 77 days to gradually reduce its altitude and attain its final orbit around Mars.

AEROCAPTURE

Aerocapture is a radical method of "putting on the brakes." If an interplanetary spacecraft is approaching a world with an atmosphere, it provides a method of decelerating for planetary capture without the use of fuel.

During an aerocapture pass, a spacecraft grazes through the outer layers of the destination world's atmosphere. During the maneuver, atmospheric drag slows the spacecraft enough to ensure capture into a planet-centered orbit. It is a precise, one-pass affair. Everything must work correctly as the probe approaches the planet on its Sun-centered orbit. It must dip into precisely determined atmospheric layers, and decelerate at one-Earth-gravity or more. If all goes well, the spacecraft emerges from the planet's atmosphere in an eccentric planet-centered orbit. The low-point of the orbit can then be raised by onboard thrusters.

The use of aerocapture can save significant spacecraft mass; which directly translates into a lower mission cost. This is done by building around the spacecraft a structure, called an aeroshell, which will protect it from the intense friction-generated heat it will experience during atmospheric entry. This aeroshell consists of a metallic or composite structure upon which is laid a thermal protection system (TPS), which will, as the name implies, provide protection to the spacecraft from the extreme heat generated during the maneuver.

Developing the "right" TPS is not trivial; and not all TPS materials work at all destinations. First of all, they come in two flavors: ablative and non-ablative. The Apollo capsule is perhaps the best-known example of a spacecraft that used an ablative TPS—one that burns away during entry. The challenge is to have enough material to ensure that it does not all burn away before the heating processes are complete, and not too much to burden the spacecraft with unnecessary weight and volume of thermal protection. The space shuttle uses a non-ablative TPS—one that does not burn up. The shuttle TPS is also designed to be reused—an expense with which missions to the outer planets on one-way trips need not be concerned.

The environment in which the TPS must operate and survive varies from planet to planet. Just as each planet is different, so are their atmospheres. Entry into the nitrogen-rich atmosphere of Earth provides a very different environment from that which would be experienced when entering the methane-laced atmosphere of Titan. Not only are the atmospheric constituents and densities different, but at Titan the interaction of the aeroshell with the atmosphere produces ultraviolet radiation to which many TPS materials are transparent! Using the wrong TPS might expose the spacecraft to mission-killing levels of radiation during the aerocapture maneuver. The heating rates, as well as the total amount of heat generated, are also different for each planet. The bottom line is that there will not likely be a single aeroshell design that will work for all missions or all destinations.

The other aspect of aerocapture that has not yet been demonstrated is the maneuver itself. The spacecraft has to hit its atmospheric destination just right, lest it bounce off into space or crash to the surface. Aerocapture is a maneuver, and the spacecraft will be flying through the atmosphere at speeds of several kilometers per second. In order to attain orbit, it must enter within a well-defined region called the "entry corridor." It is actually more like a virtual atmospheric tunnel into which the supersonic spacecraft plunges, preprogrammed to execute various maneuvers, so that it can exit into a useful orbit. If it makes too shallow an entry, it will skim off the atmosphere and go back into space—probably to be lost forever. If it comes in too steeply, the spacecraft will perform an unplanned entry and landing. (This is otherwise known as a crash!) And all of the maneuvers required to keep the vehicle within the safe corridor must be executed autonomously. The mission is likely to take place far beyond the earth, making two-way radio control impossible, and the maneuver will probably be completed before the signal of its initial entry reaches mission control back on Earth.

The first-generation aerocapture systems will probably use rigid aeroshells similar to those that have been flown for other applications. Figure 10.1 shows the nominal configuration for a first use rigid aeroshell. There is a lot known about their manufacture and performance and it only makes sense to attempt a new type of propulsion in a manner that reduces the amount of "new" technologies or systems that must be demonstrated. These systems are collectively known as rigid aeroshells, so named because they are made of hard, rigid materials. While the use of these systems will provide a tremendous mass savings on future missions, the introduction of lighter weight aerocapture systems may make them truly revolutionary. These lighter weight systems are known as ballutes. The word "ballute" comes from the combination of balloon and parachute—which describes them fairly well.

FIGURE 10.1 A prototypical rigid aeroshell as it might appear entering Mars' atmosphere during an aerocapture maneuver. (Courtesy NASA)

Ballutes are basically large, inflatable structures that provide the aerodynamic drag required to decelerate a spacecraft coming in from interplanetary space, and the lift necessary for them to maneuver into their final orbits. They are very large and allow the spacecraft to initially attain a much higher orbit than is possible using rigid aeroshells. Figure 10.2 is an artist's conception of an inflatable ballute trailing behind the spacecraft as it enters orbit around Saturn's moon, Titan. The latter reduces the overall heat loads and the former reduces further still the heat load on a given area of the ballute. When aerocapturing with a ballute, no "fireball" is evident. With a rigid aeroshell, the heat load on a square meter of aeroshell is very large due to its high-speed entry deep within a planetary atmosphere. The heat experienced by an inflatable ballute is significantly less due to its higher initial capture altitude and larger surface area (Figure 10.2).

There are attached ballutes and detached trailing ballutes. Attached ballutes look at first glance just like traditional aeroshells. They are simply inflated to encompass the spacecraft and provide the increased drag necessary for aerocapture. Trailing ballutes are just that—a large doughnut-shaped balloon trailing behind the spacecraft and attached to it by long cables.

FIGURE 10.2 Large balloon-parachutes or "ballutes" might trail behind a spacecraft allowing it to be captured into orbit around an atmosphere-bearing planet or satellite without the use of onboard propulsion. (Courtesy NASA)

Another possibility is to combine the functions of a solar sail with those of an aerobrake device. Although there are significant thermal and stress issues, at least some sail configurations are capable of withstanding high temperatures and high accelerations. Such a craft might use the sail for all post-launch interplanetary maneuvers in the inner solar system, and also to affect aerocapture.

Aerocapture Application to Solar-System Resource Surveys

Aerocapture techniques can be utilized in a number of solar-system resource-survey missions. One possibility is short-period-comet sample return.

A spacecraft could be launched from Earth and directed to rendezvous with a comet near its closest approach to the Sun. While flying in formation with the comet, the craft could retrieve samples from the comet nucleus and/or coma. After departing the comet, it could head toward Earth where aerocapture could be used to place the sample container in Earth orbit for later retrieval.

If the solar-photon sail is used as an aerobrake or in conjunction with an aerobrake device, a mission to Mars' two small satellites—Deimos and Phobos—can be conducted with minimal use of chemical propellant. In such a scenario, a probe would utilize aerocapture to decelerate from interplanetary velocities into Mars orbit. The sail would then be utilized to rendezvous successively with both of the Martian satellites. After surface samples have been collected and stored, the sail could be used to insert the probe on a trans-Earth trajectory. Aerocapture could be applied once again to slow the precious payload as it grazes Earth's atmosphere upon its return.

Outer-solar-system resources can conceptually also be surveyed with the assistance of a near-aerocapture maneuver. Distant asteroids suitably close to giant planets can be investigated if a space probe performs an aeropass in the giant's dense outer atmosphere so that it is decelerated but not captured. Such a probe will be capable of spending more time in the vicinity of selected asteroids than a non-decelerated flyby craft. Perhaps penetrator subprobes could be deployed in the upper surface layers of the asteroid as the decelerated probe passes by.

Some Aerocapture Issues

To demonstrate aerocapture feasibility for a selected planetary atmosphere, it is necessary to first develop a mathematical model of that atmosphere's density variation with height. For Earth and Mars, this is not an issue as well-understood and proven models exist for both. Thanks to the success of the Cassini/Huygens mission, Titan's atmosphere is also understood at a level that would permit aerocapture.

The next step is to determine a minimum height above the surface of the destination world for the aerocapture trajectory. This allows the calculation of parameters, including deceleration, aeroshell heating, and planet-centered velocity at the conclusion of the aeropass. Maximum stresses on the aeroshell can thus be estimated and compared with aeroshell-design limitations.

Unfortunately, a planet's atmosphere is not a static envelope. The calculated atmospheric density profile might not apply to an actual aeropass due to seasonal and regional variations. An actual aerocapture mission to another solar-system world might require onboard intelligence to compensate for variations in atmospheric density by minimum aeropass height alterations or aeroshell attitude modification. Perhaps an aeroshell could be designed that could change its shape. More likely, a spacecraft attempting aerocapture would perform last minute course alterations to compensate for whatever atmospheric conditions it finds.

AEROGRAVITY ASSIST

Aerogravity assist is an extension of the established gravity-assist maneuver (Chapter 4) with a planet or large moon to increase the speed of a spacecraft non-propulsively. Basically, aerogravity assist is a technique that would allow a spacecraft to dip closer to the planet about which it is seeking to get a boost—so close as to actually fly hypersonically through its upper atmosphere during the encounter—to increase its turning angle well above what is otherwise possible and obtain a much larger change in velocity. Currently, a spacecraft seeking a gravity-assisted boost in speed or change in direction from a planet with an atmosphere must stay well away from the atmosphere lest the fragile spacecraft experience drag and heating—resulting in a loss of the mission. If, however, it is designed with aerodynamics in mind, perhaps even having wings and the ability to fly through the upper atmosphere at many times the speed of sound, then it could get a much bigger "kick" from the planet than is currently possible. While this is certainly theoretically possible, the technologies required are nowhere close to being available today.

FURTHER READING

The application of aerobraking to several interplanetary probes is reviewed by G. Genta and M. Rycroft in *Space, The Final Frontier?* (Cambridge University Press, New York, 2003).

NASA-supported research to investigate the application of the solar sail to aerocapture is reviewed in the Appendix of G.L. Matloff's *Deep-Space Probes*, 2nd edn (Springer–Praxis, Chichester, UK, 2005). Some authors have begun to extend these preliminary calculations to the case of actual interactions between small solar sails (solar kites) and various planetary atmospheres. One of these is Andreas Thellmann, whose extensive calculations for 4.7-square-meter solar kites entering Earth's and Mars' atmospheres are included in his Diploma Thesis *Entry of a Solar Kite into a Planetary Atmosphere*, submitted to Kingston University London, Faculty of Technology, School of Engineering, in April 2005.

Some technical references about "Space Brakes" include: Hall, J.L., Noca, M.A. and Bailey, R.W., "Cost–Benefit Analysis of the Aerocapture Mission Set," *Journal of Spacecraft and Rockets*, vol. 42, no. 2, 2005; Laub, B., "New TPS Materials for Aerocapture," *Space Technology and Applications International Forum*, Albuquerque, NM, 2002; Masciarelli,

J.P., "Technology Development for Deployable Aerodynamic Develerators at Mars," *Space Technology and Applications International Forum*, Albuquerque, NM, 2002; Randolph, J.E. and McRonald, A.D., "Solar System Fast Mission Trajectories Using Aerogravity Assist," *Journal of Spacecraft and Rockets*, vol. 29, no. 2, 1992.

[See also Plate III in the color section]

11

THE ION TRAIL

Dipt in the richest tincture of the skies,
Where light disports in ever-mingling dyes,
While ev'ry beam new transient colors flings
Colors that change whene'er they wave their wings.

Alexander Pope, from *The Rape of the Lock*

THE ion drive is perhaps the most colorful of the new in-space propulsion techniques, due to photons emitted from recombined ions and electrons in the exhaust of the accelerating craft, operating ion engines emit a science-fiction-like blue glow (see Figure 11.1). It is among the most capable and versatile space propulsion systems ever devised, and, at least in its early incarnations, makes use of plentiful sunlight as its power source.

It is also one of the most venerable. Even though it is only now coming into widespread application, interplanetary propulsion by electrically charged particles has a conceptual history dating to the first decades of the twentieth century.

131

FIGURE 11.1 NASA-457 Hall thruster during testing. (Courtesy NASA; illustration also appears in L.R. Pinero and G.E. Bowers, "Multi-Kilowatt Power Module for High-power Hall Thrusters," NASA/TM-2005-213348)

ION DRIVE HISTORY

Perhaps the earliest theoretical consideration of ion propulsion was by American rocket pioneer Robert Goddard in 1906. Goddard and his students carried out some preliminary experiments in 1916. In his book, *Possibilities of Space Flight*, the German space theoretician Hermann Oberth also discussed ion propulsion, and many others expanded upon this early work during the next few decades.

Real progress, however, did not occur until after the Space Age dawned in 1957. Government contracts in the United States and elsewhere resulted in a growing industry devoted to ion-propulsion research. Laboratory tests of breadboard ion drives were performed in a number of facilities and issues of reliability and system erosion by the various propellants were addressed.

Before 1970, tests on both suborbital sounding rockets and satellites had been performed. Reliability increased to the point at which life tests demonstrated up to 8,000 hours of operation. By contrast, most chemical propulsion systems operate for seconds or, at most, minutes.

If the space race between the USA and the USSR had continued after the initial Apollo landings on the Moon, ion drives would surely have reached operational status by 1980. But the superpowers turned their attention to other arenas and initial utilization of this technology in interplanetary missions was delayed for about two decades.

ELECTRIC-PROPULSION FUNDAMENTALS

First, some terminology needs to be explained. The terms, "ion propulsion" and "electric propulsion" are often used synonymously. Electric-propulsion systems come in many flavors. The primary ones with application for deep-space missions are the gridded ion thruster and the Hall-effect thruster, both of which use ions as propellant. Both of these thrusters are operated by electrical power, generated by either sunlight falling on a spacecraft's solar arrays and being converted into electricity ("SEP," or Solar-Electric Propulsion) or by some sort of onboard nuclear power source ("NEP," or Nuclear-Electric Propulsion). SEP sees most of its application in the inner solar system, where sunlight is most intense. NEP systems, which have yet to fly operationally, use either a nuclear fission reactor or radioisotope decay to replace the solar panels as their source of electricity. This independence from the need for sunlight makes them suitable for outer solar-system applications. We will focus on SEP systems due to their source of power (inexhaustible sunlight) and their versatility for inner-solar-system science and exploration.

Figure 11.2 is a schematic representation of a gridded–ion SEP system. In both gridded ion and Hall systems, sunlight produces electricity as it strikes solar panels. A fraction of the incident sunlight (as much as 20% in state-of-the-art photovoltaic cells) is converted into electricity. The electrical energy is, after appropriate conditioning, used to ionize propellant atoms and to accelerate these positively charged ions (and negatively charged electrons) out the back of the spacecraft, producing thrust. Ion thruster efficiencies are high—as much as 60% of the electrical energy is ultimately transferred to exhaust kinetic energy. Although almost all the thrust is produced by the much more massive positive ions, electrons must be expelled to ensure that the exhaust is electrically neutral and electric charge does not build up on ion engine surfaces.

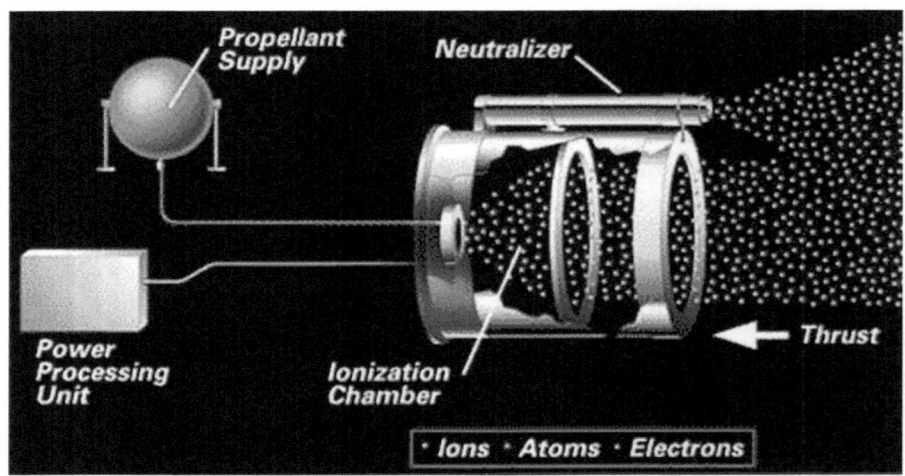

FIGURE 11.2 A gridded-ion SEP. (Courtesy NASA)

Ion drives may be low thrust, but their exhaust velocities are high—30 kilometers per second or higher for state-of-the-art thrusters. Typical accelerations are in the range of 0.0001 of an Earth-surface gravity, or about one millimeter per square second. Electric propulsion is therefore inadequate for Earth-to-orbit transportation, but it is ideally suited for use in deep space. In space, the small, but continuous thrust will accelerate a spacecraft to higher and higher velocities over long periods of time. And when you are traveling between planets, no matter what you do, you have a lot of time!

Research currently underway will result in increased exhaust velocity, thruster efficiency, solar-panel conversion efficiency and decreased solar-panel mass. Exhaust velocities in excess of 100 kilometers per second are certainly possible.

In an ion thruster, ionized fuel atoms and electrons are accelerated by an electric or magnetic field. Typical fuels are those that are readily ionized and easy to store. Candidate fuels are argon, krypton, cesium, mercury and xenon. The current preferred choice is xenon because it is non-toxic and easy to store. Market speculators beware: widespread use of SEP (or NEP) will tax the current world availability of xenon.

Modern SEP engines can operate at power levels as high as 50 kilowatts—which is fine for robotic probes, but is not yet at the megawatt level required to propel peopled interplanetary missions. The power-to-mass ratio of state-of-the-art ion thrusters is in the range of hundreds of watts per kilogram.

Hall thrusters are pursuing a parallel development path, and have had their own string of technical and mission successes. Though studied in the United States as early as the 1960s, serious Hall thruster development occurred in the former Soviet Union almost exclusively until the early 1990s. Hall thrusters, like gridded ion thrusters, accelerate ions by using electric fields. However, the orientation of the field and the absence of acceleration grids that give gridded ion thrusters their name, allow Hall-effect thruster systems to have fewer parts and, by comparison, greater simplicity. They are currently in use on board Earth-orbiting spacecraft where they provide low fuel consumption station-keeping and orbital maneuvering capability.

A technology closely related to the ion drive is the plasma rocket. In the plasma rocket, a magnetic field is used to accelerate an electrically neutral but ionized fluid. Although plasma thrusters are less developed than electric thrusters, they show promise of raising engine thrust, albeit with somewhat reduced exhaust velocities.

INITIAL INTERPLANETARY APPLICATION OF ION PROPULSION

A peopled mission using ion propulsion to escape LEO for an inner-solar-system destination might not be practical. The astronauts might accumulate a dangerously high radiation dose as their craft slowly spirals up through Earth's Van Allen radiation belts. However, SEP may be applied to peopled interplanetary missions departing from higher orbits.

As of this writing, several interplanetary probes have been launched utilizing SEP for post-launch, or post-Earth-escape prime propulsion. The first of these was NASA's Deep Space 1 (Figure 11.3), which was launched from Cape Canaveral in October 1998 on board a Delta 2 rocket.

Deep Space 1, with a launch mass of 489 kilograms, was propelled after Earth escape by a xenon-fueled solar-electric rocket with an exhaust velocity of about 30 kilometers per second. As part of the NASA New Millennium Program, Deep Space 1 was primarily a technology demonstrator of 12 new technologies, including SEP.

The solar arrays of the probe provided a maximum of 2.4 kilowatts of electricity, which was used to power scientific items including imagers, particle monitors, and an infrared spectrometer as well as the ion engine.

After Earth escape, the ion drive was turned on. The SEP on board Deep Space 1 operated successfully for several hundred days in space. The drive was operated continually for 2.5 months, expelling about 20

FIGURE 11.3 NASA's Deep Space 1 approaches comet Borrelly. (Courtesy NASA)

kilograms of xenon propellant and altering the craft's heliocentric velocity by approximately 1 kilometer per second.

During its three-year mission, Deep Space 1 was redirected to approach several inner-solar-system objects. It returned excellent images of Comet Borrelly and Asteroid Braille during its close-up exploration of these bodies.

After the successful demonstration of SEP technology on board Deep Space 1, other ion-propelled probes were launched in rapid succession. Europe's SMART-1 demonstrated the utility of Hall thrusters in performing long-duration Earth-escape maneuvers. This probe required approximately 13 months to spiral from LEO to the vicinity of the Moon. Although much more time was required for cis-lunar maneuvers than would have been necessary by chemical rockets, ion propulsion is much more fuel-efficient.

Another SEP-propelled interplanetary venture, Japan's Hayabusa (Muses-C), was launched in May 2003. This craft is currently performing a sample return mission to Asteroid 25143 Itokawa. If all continues to go well, samples should parachute to Earth near Woomera, Australia, in June 2007.

POSSIBLE ION PROPULSION TECHNOLOGY APPLICATION TO SOLAR-SYSTEM DEVELOPMENT

So how will ion propulsion help us to "live off the land" and explore the solar system? Because of their inherently high efficiency, 10 times greater than their chemical propulsion cousins, they will allow more complex, propulsion-intense missions than would be otherwise possible. They are powered by the Sun, through the conversion of sunlight into electricity via their photovoltaic arrays. Lastly, their propellant should be able to be extracted from the atmospheres of many planets and moons throughout the solar system. Noble gases, which are the propellants of choice for ion drives, are not uncommon in the solar system. Comets, once thought to be devoid of argon and other noble gases, were found in the year 2000 to contain them by a team from the Southwest Research Institute using sounding rockets to study the comet Hale-Bopp. Argon, krypton, and xenon—all being potential fuel for ion drives—were also found at Jupiter by NASA's Galileo mission in 1996. (Actually, the data was collected in 1996, but the discovery was not announced until 2000, after the data was more thoroughly analyzed.)

The technology of the ion rocket has many applications to space development. Slow, but highly efficient SEP tugs could be used to transfer cargo between the Earth and the Moon, Mars or other inner-solar-system destinations. Without human beings, who require relatively rapid interplanetary trip times due to their susceptibility to the hazards of weightlessness and space radiation, cargo-carrying SEP tugs can ply the space between worlds slowly and efficiently. Why is this of such benefit? Simply put, the impact of their high efficiency ripples through the mission design and planning, and results not only in significant cost savings but also in much lower overall mission mass. Recall that SEP systems are approximately 10 times more efficient than their chemical cousins. This means that they will use approximately 10 times less fuel to get to the same destination as a chemical rocket. Ten times less fuel must therefore be launched into space—significantly reducing the overall cost of launch. If they can be refueled at their destination, as recent data suggests might be possible, then they can be reused without resupply from the home planet, further increasing their cost benefit for tomorrow's explorers.

This mass benefit and resultant cost savings make SEP systems attractive for human exploration, robotic science, and industrial development of space. Some satellites in orbit around the Earth use Hall and gridded ion thrusters today. Communications satellites, whose profit-making potential is closely tied with both on-orbit lifetime and the number of transponders on board, will benefit from SEP. They can achieve longer life by using SEP to perform in-space maneuvers, instead of chemical thrusters which would more quickly run out of fuel. They can also launch more transponders, adding useful payload mass to the satellite, by offloading propellant for fuel-hungry chemical propulsion systems and replacing it with fuel-efficient ion propulsion.

SEP is the next logical step in rocket propulsion. It is a technology seeking worldwide application, with successful missions now having been launched by Russia, ESA, Japan and the United States. Many more missions will soon follow, opening the ion trail for a sustainable exploration of the solar system.

FURTHER READING

One source outlining the early history of ion-drive concepts is George R. Brewer's *Ion Propulsion* (Gordon & Breach, New York, 1970). A more up-to-date treatment of electric propulsion and plasma rocket fundamentals can be found in Martin J.L. Turner's *Rocket and Spacecraft Propulsion*, 2nd edn (Springer–Praxis, Chichester, UK, 2005). Early operational missions using this technology are discussed by Giancarlo Genta and Michael Rycroft in *Space: The Final Frontier* (Cambridge University Press, New York, 2003).

[See also Plate IV in the color section]

12

THE ORBITAL STEAM LOCOMOTIVE

Put in your water and shovel in your coal.
Put your head out the window, watch them drivers roll!
I'll run her till she leaves the rail,
Cause we're eight hours late with the western mail!

From the US folksong *Casey Jones*,
attributed to Newton and Seibert (1909)

T HE development of the American west, to a very large extent, depended upon the steam locomotive. Not only did the coal-driven trains of the transcontinental railroad transport the mail, but such items as tools, produce, and people were carried by this service.

Long before Casey Jones, and long before the taming of steam, people had investigated the physics of the steam engine. In fact, as briefly discussed in Chapter 3, demonstrations of thrust produced by steam escaping from a vent may be the first historical application of the rocket principle.

In 50 BC, Hero of Alexandria constructed an "aeropile," which consisted of a boiler and two vertical pipes that were attached to a horizontal axle on which a hollow sphere was mounted. A vent was located within the sphere.

The boiler was filled with water. A fire was lit under the boiler; steam rose into the hollow sphere and was forced out of the vent. The reaction to this "thrust" caused the aeropile to spin, much to the delight of onlookers.

Perhaps, if this device had been productively employed in the Greco-Roman world, the Industrial Revolution would have dawned 1,500 years earlier! But alas, the application of the aeropile might have reduced the need for menial labor, and slavery was a sociologically important institution in the classical world.

Many centuries after Hero, engineers realized that the vented steam could be impacted against the blades of a turbine, and so the steamboat was born, in which the spinning turbine was itself attached to a paddlewheel. These water craft successfully competed with and ultimately replaced sailing ships. In the nineteenth century, the spinning turbine was attached to a wheel axle, miles of track were laid and the first coal-fueled locomotives began to chug across the landscape. Today, a new type of steamship is being considered. This one uses a very different type of steam and will travel between planets, not just across a continent. It is called a solar-thermal rocket.

SOLAR-THERMAL ROCKET FUNDAMENTALS

The operation of a solar-thermal rocket is presented in Figure 12.1. Unlike a chemical rocket, the Sun powers solar-thermal rockets. Using either a Fresnel lens or a parabolic mirror, sunlight is concentrated to superheat propellant, which is then vented to produce thrust. The concentrator is the key feature that distinguishes this technology from other propulsion technologies, heating the propellant with the equivalent of up to 1,000 Suns (at 1 AU). Recall that exhaust velocity and thrust are closely related to the propellant's temperature—and a solar-thermal system can make the propellant very, very hot.

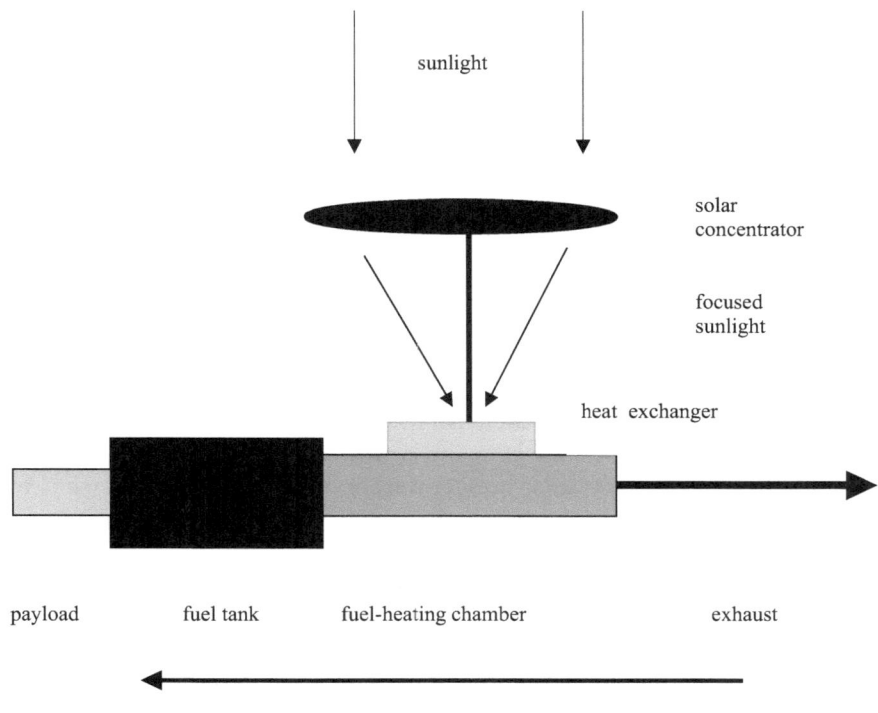

FIGURE 12.1 *The solar-thermal rocket*

Although very simple in concept, implementation is not so straight-forward. One technical challenge is the heat exchange with the propellant. As one cannot directly heat a moving fluid in a vacuum, it must be done indirectly. In indirect heat exchange, the sunlight warms a material, or "heat exchanger," that subsequently transfers the heat to the propellant.

The fuel-heating chamber must be constructed of high-temperature materials. For optimum performance, propellant temperatures are well in excess of 1,000 degrees Celsius.

Although the working fluid (propellant) in Hero's device, steamboats and coal-stoked locomotives was water, optimum performance for solar-thermal rockets is achieved if a lower molecular mass propellant is used. Current prototype solar-thermal rockets use hydrogen propellant. If we replace hydrogen with water propellant, the exhaust velocity would decrease from about 10 kilometers per second to about 3 kilometers per second.

Although the use of hydrogen as a propellant results in high exhaust velocities (typically about twice those of the best chemical rockets) there are significant drawbacks to employing this fuel on long-duration space missions. Liquid hydrogen is difficult to store in the space environment and, due to its very low boiling point of −252 degrees Centigrade, it tends to evaporate or "boil off." Even the cold temperatures of space are too warm to prevent hydrogen from turning into a gas. Hydrogen gas is notoriously difficult to store. Its low molecular weight and atomic size allows it to slip through very small cracks, making it virtually impossible to completely contain. As an alternative, some solar-thermal rocket systems are being designed to operate using methane, paying the performance penalty associated with its higher molecular weight and subsequent lower exhaust velocities, in preference to the long-term storage problems posed by hydrogen fuel.

In terms of thrust, solar-thermal rockets are intermediate in performance between chemical rockets and ion rockets. In part, because of the requirement for massive solar concentrators, no solar-thermal rocket will ever lift off from a planetary surface. But accelerations of 0.01 Earth gravity are possible in the space environment.

NEAR-TERM APPLICATION OF SOLAR-THERMAL ROCKETS

NASA, the US Department of Defense, and other agencies are considering near-future applications of this technology. Perhaps the earliest application will be orbit transfer.

Assume that you've paid the launch cost to put a satellite into low Earth orbit a few hundred kilometers above Earth's surface and you wish to raise its orbital height to a geosynchronous altitude, about 36,000 kilometers above Earth's surface.

You could of course do this with a conventional upper stage using chemical propulsion, but a hydrogen-expelling solar-thermal rocket has about twice the efficiency of the best chemical rocket. So there is a significant economy in developing a solar-thermal tug to loft payloads between low Earth and geosynchronous orbits. This economy will be of interest to developers of communication, navigation, and Earth-viewing satellites.

But, as generally happens in the space business, there are trade-offs. A chemical rocket is a high-thrust device and the orbit transfer can be accomplished within hours or, at most, a few days. Although a solar-thermal rocket has a higher thrust than a solar-electric rocket, the orbit-transfer time between low Earth orbit and geosynchronous orbit is considerably longer than for a chemical rocket.

Orbital-transfer tugs based upon the solar-thermal rocket (or the solar-electric rocket) are certainly feasible. But payloads on board these tugs will require more shielding during their transfer through Earth's Van Allen radiation belts than payloads on board chemical rocket orbital-transfer tugs due to the fact that they will be spending more time in the radiation belts, increasing the total radiation exposure and potentially causing more radiation-induced damage.

At least two flight tests of the technology were proposed in the 1990s. The Shooting Star Experiment was to have launched from the space shuttle orbiter and demonstrate the fundamentals of the technology. An artist concept for the Shooting Star is shown in Figure 12.2. Boeing, working with the US Air Force, was tasked to develop a space tug using solar-thermal propulsion called the "Solar Orbital Transfer Vehicle (SOTV)." Neither the Shooting Star Experiment nor the SOTV flew.

FIGURE 12.2 Artist's concept of NASA's shooting star experiment. (Courtesy NASA)

POSSIBLE APPLICATION OF SOLAR-THERMAL TECHNOLOGY TO SOLAR-SYSTEM DEVELOPMENT

Solar-thermal rockets are ideally suited for "living off the land" in space. With sunlight as their source of energy and abundant hydrogen or methane as their fuel, they can operate anywhere within the orbit of Mars with relatively high thrust and high efficiency—a compromise between the best (and worst) of both chemical and electric propulsion systems. Departing the Earth with cargo, or returning to it with raw materials, solar-thermal-propelled spacecraft can carry large payloads fairly quickly and efficiently.

They can be refueled at comets, whose abundant water can be cracked into hydrogen and oxygen by passing through the water an electrical current and collecting the liberated gases in a process known as "hydrolysis." This might also be implemented at the Moon where water ice is thought to exist in forever-shadowed craters near the poles.

As well as its potential utilization as a space drive to shunt freight around the solar system, solar-thermal technology may be of use to future space-mining processes and industrialization enterprises.

The solar concentrators for a solar-thermal interorbit tug capable of ferrying large payloads will themselves be large. In one design described by Robert Salkeld and his collaborators, a 100-meter collector diameter is proposed. So a solar-thermal solar concentrator will have to be a low-mass item, with its thickness measured in microns. In order to heat the working fluid to the requisite high temperatures, the concentrator must also be precisely machined and capable of withstanding the space environment for long time periods. Near-term small satellite missions using solar-thermal propulsion would not have to be nearly as ambitious. Concentrators a few meters across are sufficient to produce thrust for missions in this class.

These large-scale concentrator requirements will be of great interest to those who would mine small solar-system bodies for water, other volatile substances, and higher melting point compounds or elements. A space miner might simply use the solar-thermal concentrator to focus sunlight on his or her asteroid and collect pure materials as they boil off.

FURTHER READING

Hero's experimental aeropile is described by Carsbie C. Adams in his classic *Space Flight* (McGraw-Hill, New York, 1958). Some even earlier experiments are mentioned by Eugen Sanger in *Space Flight* (McGraw-Hill, New York, 1965).

The relationship between solar-thermal (and nuclear-thermal) rocket exhaust velocity and propellant molecular mass is discussed by Martin J.L. Turner in *Rocket and Spacecraft Propulsion*, 2nd edn (Spinger–Praxis, Chichester, UK, 2005).

An early design for a large solar-thermal rocket is included in *Space Transportation Systems*, which was edited by R. Salkeld, D.W. Patterson and J. Grey and published by AIAA Press (Washington, DC, 1978). A much more recent consideration of the utility of solar-thermal solar concentrators in space mining is included in J.S. Lewis' *Mining the Sky* (Addison-Wesley, Reading, MA, 1996).

[See also Plate V in the color section]

13

SKY CLIPPERS

I must go down to the sea again, to the lonely sea and the sky,
And all I ask is a tall ship and a star to steer her by
And the wheel's kick and the wind's song and the white sail's shaking,
And a grey mist on the sea's face and a grey dawn breaking.

John Masefield, from *Sea Fever*

IT was before the dawn of recorded history that the first great Age of Sail began. Perhaps along the Nile River, a genius lived who observed the ways of swans and reached a brilliant conclusion. If birds could fluff out their feathers to catch the wind, and drift effortlessly against the current, why couldn't humans learn the same trick?

Soon, crude sailboats were plying the navigable reach of that great river; and somewhat more sophisticated craft were beginning to cross the Mediterranean toward the fertile Cycladic Islands.

Eventually the designers of sailing craft grew bolder and more confident. Multiple sails and keels were employed and the white sails caught the wind's song as far afield as the Pillars of Hercules and the British Isles.

149

Europeans would never have settled the Americas without the Tall Ship. In its ultimate incarnation—the nineteenth-century Clipper, the sailing craft enabled a globe-circling trading network. Before the creation of the Transcontinental Railroad and the Panama Canal, ships sailed through the treacherous waters south of Tierra del Fuego in Argentina and proceeded up the Pacific coast to visit new settlements in the exotic land called California.

There is much romance in the exploits of the sailors as they visited far lands and encountered exotic peoples and strange creatures. It is most exciting that a second great Age of Sail seems to be dawning, as researchers experiment with spacecraft that can sail the Sun's photon breeze.

PHOTON SAILING HISTORY

Solar sailing history begins in 1873, when James Clerk Maxwell demonstrated that the photon, although without mass, has momentum. For the first time, astronomers could understand why a comet's tail always points away from the Sun.

In 1900, the Russian physicist Peter Lebedew experimentally measured photon radiation pressure. In the early twentieth century, physicists including Albert Einstein incorporated photon momentum into quantum mechanics.

However, two Russians were the first to realize that the pressure of sunlight could be used to propel a thin-film spacecraft through the cosmos. These pioneers, Konstantin Tsiolkovsky and Frederick Tsander, developed the preliminary theory of solar sailing in the 1920s. The theory of operation is beautiful in its simplicity. Sunlight, composed of many individual particles, called photons, strike the lightweight, highly reflective sail and reflect back in roughly the opposite direction from which they came. Their momentum is imparted to the sail during the reflection, producing a force that ever so slightly moves the sail forward. The process is conceptually similar to an Earth-wind propelled sail, hence the name, "solar sail."

Surprisingly, not much happened with the concept until the 1950s when American researchers, including Carl Wiley and Richard Garwin began to investigate how one would apply solar sailing to interplanetary travel. In 1959, T.C. Tsu published an analysis of optimized interplanetary trajectories using solar sails.

During the USA–USSR Space Race, high-thrust devices capable of achieving Earth orbit and reaching the Moon were given developmental priority. Low-thrust, in-space propulsion systems such as the solar sail had to wait their turn.

Solar-radiation pressure was a significant orbital perturbation during the flights, in the 1960s, of the large orbital balloons Echo 1 and 2. The first operational application of solar sailing occurred during the 1970s, when the controllers of Mariner 10, the first spacecraft to perform multiple flybys of Mercury, could conserve steering fuel by using solar radiation pressure on the craft's solar panels for some attitude adjustments.

After the success of this mission, managers at NASA's Jet Propulsion Laboratory realized that solar sails might have many inner-solar-system applications. One of the options they considered for the 1986 Halley comet rendezvous was solar sailing. A solar sail unfurled near the Sun was also considered for propulsion of another 1980s NASA probe study, the TAU extrasolar probe to 1,000 Astronomical Units.

But these were paper studies. It was still necessary to demonstrate that large, micron-thin structures could be unfurled and controlled in the zero-gravity space environment. The first successful in-space unfurlment of a sail-like structure occurred in 1993, when a 20-meter-diameter spinning thin-film reflector called Znayma was unfurled from a Progress ferry vehicle visiting the Mir space station. In 1996, an American space shuttle deployed an inflatable thin-film radio antenna in space.

In the summer of 2004, the Japanese space agency succeeded in unfurling in space two solar sails that had been folded for launch using the principles of origami. This successful suborbital test was followed in 2005 by an attempt to orbit a solar-sail spacecraft. Sadly, Cosmos-1, funded by The Planetary Society, was lost because of the failure of its Russian booster.

Unfurlment tests of increasingly larger sails have continued in ground-based vacuum chambers. NASA hopes to demonstrate sail deployment techniques in the near future by unfurling a 100-meter sail in a domed football stadium.

SOLAR SAILING FUNDAMENTALS

As is true for aerocapture and the electrodynamic tether, the solar photon sail is an example of propellantless field propulsion. Thrust is produced not by the expelling of onboard fuel, but by an external force field—in this case the pressure of solar photons striking the sail. As such, application of the sail can only produce acceleration within the inner solar system or in the presence of another copious source of unidirectional photons.

If we assume a highly reflective, non-transmissive sail with a reflectivity to sunlight of R_{sail}, and if we discount the small contribution to acceleration by photons absorbed by the sail, spacecraft acceleration due to solar radiation pressure can be approximately calculated (in meters per square second) as:

$$ACC_{sail} = \frac{1 + R_{sail}}{cM_{s/c}} S_c A_{sail}$$

where c is the speed of light (300,000 kilometers per second), $M_{s/c}$ is the total spacecraft mass in kilograms, S_c is the solar flux (about 1,400 watts per square meter at the Earth's distance from the Sun), and A_{sail} is the sail's area normal to the Sun in square meters. Solar photon flux decreases with the square of solar distance. At twice the Earth's solar separation (2 Astronomical Units), S_c has fallen to about 350 watts per square meter, or one-quarter of its previous value.

The phenomenon which describes this decrease in sunlight intensity is known as the "inverse square law," and is common in many physical processes involving a point source of radiation (like the Sun). Figure 13.1

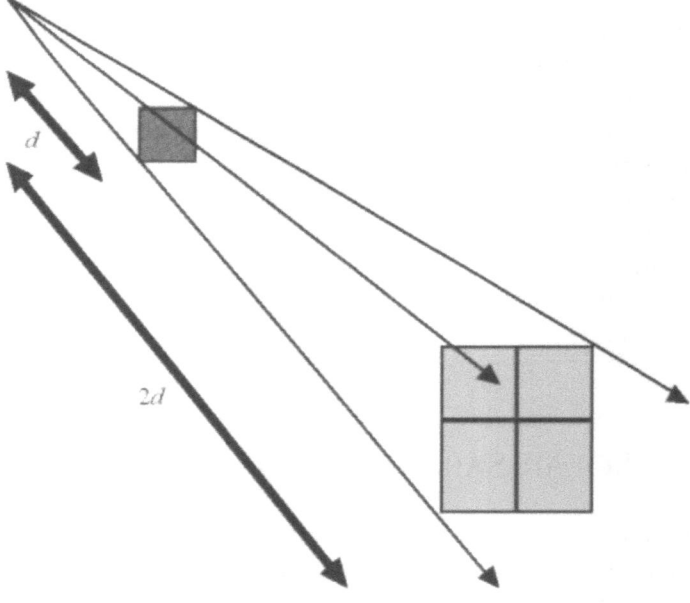

Intensity = $1/d^2$

FIGURE 13.1 The inverse square law demonstrated.

illustrates how the inverse square law works. As the distance from the Sun to the solar sail increases, the amount of light spreads out over a larger and larger area, and the light falling on the sail becomes correspondingly less intense. If you were to double the distance between the solar sail and the Sun, the amount of light falling on the sail would only be one-quarter of its previous value—4 is 2 squared and is in the denominator of the fraction, hence the term "inverse square." Similarly, if you were to move the sail to a distance 4 times further from the sun, the amount of sunlight falling on the sail would be down to one-sixteenth of its original value.

To maximize sail acceleration, it is necessary to minimize spacecraft mass, maximize sail reflectivity and work with the largest possible sail areas. Sail engineers often characterize sail performance using a parameter called the "lightness factor." The lightness factor is the ratio of solar radiation-pressure force on the sail to solar gravitational force.

Since solar gravitational force also decreases with the square of the solar distance, lightness factor is constant for a sailcraft anywhere within the solar system. One interesting result can be obtained by assuming a lightness factor of exactly 1, for a sail unfurled normal to the Sun near the Earth at a solar distance of 1 Astronomical Unit (150 million kilometers). Application of Newton's Laws of Motion reveals that this craft moves off into space in a straight line at a constant velocity of 30 kilometers per second. Such a craft could reach Mars in about a month. But astronauts would then have to address the "small" question of how to slow down!

CURRENT SAIL TECHNOLOGY

At present, all solar sails are carefully folded after construction, placed in a rocket's upper stage, and unfurled in space. Sail films are generally tri-layered. A highly refective aluminum layer (which faces the Sun) is affixed to a plastic substrate. One commonly used substrate is Kapton^{TM}. An emissive layer, often chromium, is deposited on the side of the sail directed away from the Sun. Its function is to radiate the heat produced by the small amount of solar flux absorbed by the sail's aluminum front face.

Today's "first-generation" sails are very thin—typically a few microns in thickness. Sail areal mass thickness is in the vicinity of a few grams per square meter. You would obtain a film of this mass if you could squash a few raisins and spread them evenly across your table top! Various structural fittings—beams, spars, and cables—add 30% or more to a solar photon

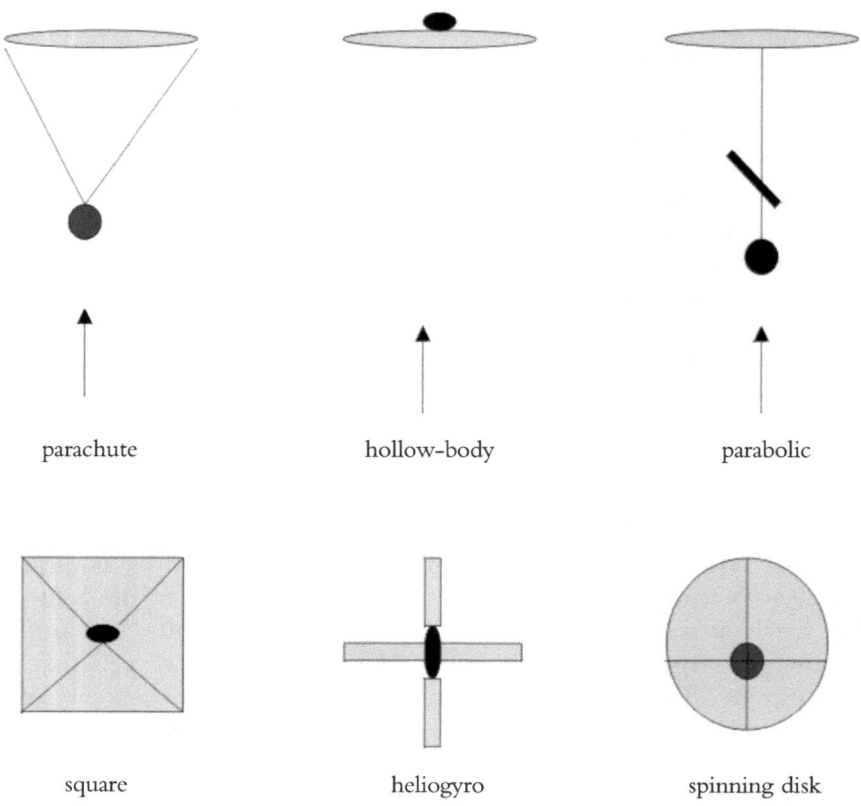

parachute   hollow-body   parabolic

square   heliogyro   spinning disk

FIGURE 13.2 Six solar-photon-sail configurations. Cables and spars are solid lines. Payloads are represented by small solid ovals. Arrows denote sunlight and spacecraft acceleration direction.

sail's mass. Experiments have also been performed with inflatable structural elements.

A number of sail configurations have been suggested, researched, or flown. Some of these are shown in Figure 13.2.

In the parachute sail, sunlight pushes against a sail canopy. Payload is attached to the sail by high-tensile strength cables.

The hollow-body, or pillow, sail is inflatable. Sunlight pushes against the "lower" sail surface, which supports the payload.

A more complex arrangement is the parabolic sail, also called the solar-photon thruster (SPT). Here, sunlight is incident against a large, curved "collector" sail and focused against a small thruster sail. The SPT allows for the possibility of "tacking" at a larger angle against the direction of the photon stream.

In a square (or rectangular) arrangement, the payload is attached with spars to the center of the structure. It is this technology that has been the focus of NASA's recent investments.

NASA recently completed a three-year effort to demonstrate key solar sail technologies from two competing teams. L'Garde Inc., of Tustin, California, developed a solar sail system that employs booms that are flexible at ambient temperatures but "rigidized" at low temperatures. The L'Garde sail uses articulated vanes located at the corners of the square to control the solar sail attitude and thrust direction (Figure 13.3). ATK of Goleta, California, developed a coilable longeron that deploys in space in much the way that a spring-loaded screw is rotated to remove it from an object. Once deployed, the sail is unfurled using a network of cables—in a method analogous to the sailing ships of yesterday. Both hardware vendors fabricated and tested 10-meter subscale solar sails in the spring of 2004; and 20-meter subscale solar sail deployments were conducted under thermal vacuum conditions in 2005.

A spinning heliogyro has four or more sail blades and spins in the direction normal to sunlight, like a gyroscope. The spinning-disk sail is also stabilized by centrifugal force. Its payload is mounted at the center of the sail and supported by a network of beams and spars.

FIGURE 13.3 The L'Garde solar sail, photographed prior to thermal/vacuum testing by NASA. (Courtesy NASA)

There are many possible variations on sail configuration. The hoop sail, for instance, combines aspects of the parachute and spinning-disk sails. In a hoop sail, structural support takes the form of a hoop the same radius as the sail film and concentric with the sail. Distributed payload components could be suspended from the hoop.

All of the configurations shown in Figure 13.2 have advantages and disadvantages. For instance, the inflatable hollow-body sail is perhaps the easiest to deploy; however, it is also very prone to micrometeorite damage that might release inflating gas during periods of high acceleration.

Solar-photon sails will be huge by any measure. Interplanetary robotic sails have projected diameters of 40–400 meters. Sails supporting human exploration of Mars will carry heavier payloads and will therefore have diameters in the 1–10 kilometer range. Ultimate, space-manufactured sails used to divert asteroids or for interstellar travel could be 100 kilometers in diameter or larger.

MISSIONS FOR NEAR-TERM SOLAR-PHOTON SAILS

Because of the inverse-square nature of solar radiant flux, the solar-photon sail will accelerate fastest in the inner solar system. Intuition might suggest that an Earth-launched sail could never closely approach the Sun because the solar flux is directed radially outward from the Sun. But in this case, intuition is wrong for the following reason.

Everything in the solar system is in some sort of orbit and therefore, thanks to Mr Newton, we know that they will continue to move with the same orbital energy unless acted upon by some outside force. A sailcraft leaving the Earth would still be in orbit around the Sun. It would not "fall" toward the Sun unless some force acts upon it to change its velocity. A sail could be used to reflect sunlight in such a way as to slow the spacecraft in its orbit, thus causing it to spiral in toward the Sun. Once the new, presumably desired orbital distance from the Sun has been achieved, the sail could be reoriented to produce thrust in the proper direction in order to allow the craft to remain in this new orbit or further propel it to another location.

As presented in Figure 13.4, the sail can spiral either closer to the Sun or farther out in the solar system, depending upon its tilt angle (also called the aspect angle).

Perhaps the premier scientific mission for the solar-photon sail will be station-keeping for solar observatories. For many years, solar-storm early warning satellites, such as NASA's Advanced Composition Explorer

(ACE), have been permanently stationed 1.5 million kilometers closer to the Sun than the Earth's orbit. The main function of these craft is to warn of massive solar flares directed toward Earth that have the potential of increasing orbital radiation levels and disrupting intercontinental communication. The lifetime of solar observatories is limited by the requirement to carry maneuvering fuel. The solar-photon sail could be used for all post-Earth-escape maneuvers, reducing mass and cost and increasing operational lifetime.

If you construct a three-dimensional drawing in the manner of Figure 13.4, you'll readily see that the solar-photon sail can also direct a component of reflected solar flux perpendicular to the ecliptic, the plane of the Earth's solar orbit. Because of this capability, solar sails are also capable of propelling extra-ecliptic missions such as solar-polar observatories and comet rendezvous probes.

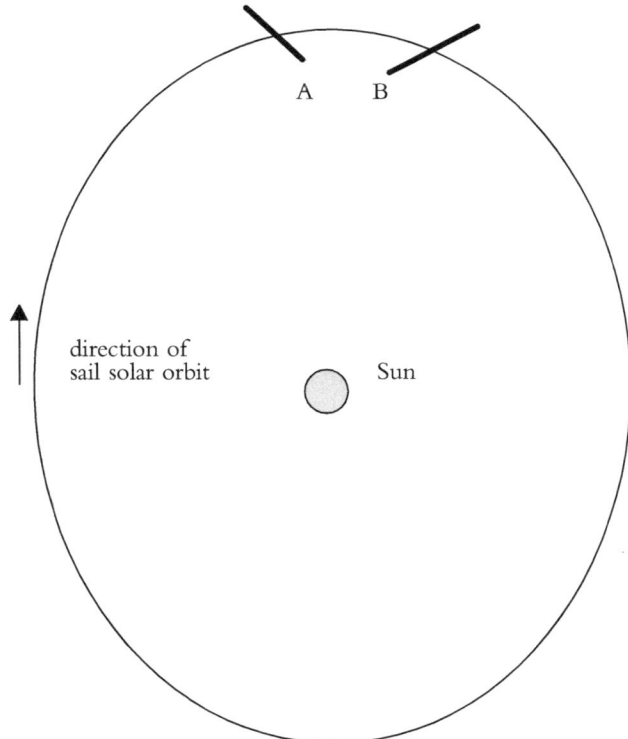

FIGURE 13.4 Two solar sails (A and B) orbiting the Sun. Because solar radiation pressure on sail A produces a thrust component tangent to and in the same direction as sail A's orbit, sail A spirals outward. Sail B will spiral inward toward the Sun because the solar radiation thrust component opposes the orbital direction.

Solar-photon sails in a number of configurations also have advantages for designers of communication, navigation, and Earth-observation satellites. It is possible to design a sail capable of maintaining a more-or-less constant high-latitude position analogous to an equatorial geosynchronous satellite, but at a greater distance from the Earth.

Most sail designers expect that atmospheric drag might limit sail operations to orbital heights greater than 1,000 kilometers. But it is not impossible that parabolic-rigged sails could utilize Earth-reflection to unfurl at lower altitudes.

Atmospheric drag on sails might actually be a good thing. Analysis indicates that certain sail configurations could withstand accelerations greater than one Earth gravity. Sails could therefore serve as aerocapture devices.

But for the present, sails, like ion drives, will be low acceleration devices. Months or years must be devoted to the propulsive phase of any scientific mission propelled by the solar-photon sail.

NEAR-FUTURE SOLAR-SAIL DEVELOPMENT POSSIBILITIES

A number of possibilities are under investigation that promise to improve sail performance and versatility. Hyper-thin sail films could be constructed by replacing the plastic sail substrate with a material that would degrade and evaporate when exposed to solar ultraviolet radiation. A monolayer sail constructed from a mesh of carbon filaments could conceptually be launched from Earth and unfurled in space to replace the much more massive tri-layer sails currently under development.

One mission under study for the 2020 time frame is an extrasolar probe. First injected into a near-Sun near-parabolic orbit using a Jupiter flyby or near-Earth orbit manipulations, the sail would be unfurled at a perihelion of 0.2–0.3 AU and accelerated out of the solar system. (Please refer to Chapter 8 for more information on near-term interstellar missions.)

Solar-system exit velocities using present-day sail designs could approximate 10 AU per year, about three times the velocity of the Voyagers. A 30-kilogram science payload on board a 250-kilogram sailcraft with a football-stadium sized sail could reach the limits of the heliosphere and explore near galactic space at 200 AU from the Sun, after a flight of "only" 20 years. But current technology sails require more than 10,000 years to cross the interstellar gulf between the Sun and Alpha

Centauri. Technological improvements are possible that might some day reduce this crossing time to a millennium or less.

SOLAR-PHOTON SAILS AND SPACE DEVELOPMENT

If you want to carry freight across the solar system, the sail is the transit system of choice. Using the Sun for energy and propulsion, sails could be used to carry habitats and landing craft to Mars with minimal cost, though at slower velocities than the smaller human-occupied craft that would follow them.

Sails could divert asteroids and comets targeting the Earth, if impact-warning times are measured in decades. During the diversion process, space miners could also transfer valuable resources from these errant objects to space fabrication facilities close to the Earth using solar-sail space "clippers." Some of these freighters will be large enough to be viewed easily through small, terrestrial telescopes. The exploits of these sailing ships of the cosmos will inspire generations of future space pioneers.

FURTHER READING

A number of good sources on solar-sailing fundamentals and history have been published. A very popular and accessible reference is Louis Friedman's *Starsailing* (Wiley, New York, 1988). For more of the engineering details, consult Jerome L. Wright's *Space Sailing* (Gordon & Breach, Philadelphia, PA, 1992). A more up-to-date, and more mathematical, treatment is Colin R. McInnes' *Solar Sailing* (Springer–Praxis, Chichester, UK, 1999).

Many representations of solar-sail acceleration have been published and all derive from Einstein's work in the early twentieth century. The solar-sail acceleration equation in this chapter derives from Chapter 4 of G.L. Matloff's *Deep-Space Probes*, 2nd edn (Springer–Praxis, Chichester, UK, 2005).

Various solar-photon sail configurations and missions are further discussed in the references listed above.

14

ART OR SCIENCE?

Since I can never see your face,
And never shake you by the hand,
I send my soul through time and space
To greet you. You will understand.

James Elroy Flecker, from *To a Poet a Thousand Years Hence*

THE Interstellar Message Plaques, affixed to our tiny Pioneers and Voyagers as they cruise toward the boundaries of nearby interstellar space, are messages to the far future. They have been compared to letters sealed in bottles and dropped in the ocean, in the hope that they might be retrieved and read in distant climes, on far shores.

Future interstellar probes will certainly follow the tradition established in the 1970s by the designers of Pioneers 10 and 11 and Voyagers 1 and 2. Some form of plaque will be mounted on the spacecraft to speak for Earth and humanity, in case spacefaring extraterrestrials encounter the probe in the interstellar vastness. But even the most advanced galactic space travelers might not be able to classify the message plaque as a work of art or a work of science.

Even if ET could make that connection, it is very doubtful that he or she could understand one fascinating and distinctly human aspect of message-plaque design—the fascinating controversy elicited by our species' first calling cards to the universe.

THE MESSAGE PLAQUES

The Pioneer Plaques

As mission planners prepared Pioneer 10 for its 1972 launch toward Jupiter and its sister Pioneer 11 for its flyby of Jupiter and Saturn, they realized the significance of these small probes in terrestrial history. Both Pioneers would gain sufficient orbital energy during their giant-planet encounters to ultimately leave the solar system—to become humanity's first interstellar spacecraft.

Because there is a small probability that technological extraterrestrials might intercept one of these craft during the next billion years or so, a small team designed a visual message to intelligent aliens—a plaque that could be interpreted by anyone with eyes and technology. Team members on this project included a planetary scientist (Carl Sagan), a radio astronomer (Frank Drake) and an artist (Linda Salzman Sagan).

The final product, which was affixed to each of the pioneers, was a 15 by 22.5-centimeter gold anodized aluminum plate (shown in Figure 14.1). Lots of information is encoded in this plaque.

Note the two nude human figures. Their features incorporate characteristics common to most of the human race. The male is a bit taller than the female, which is generally true for humans. His right hand is lifted in a typical human gesture of greeting or good will.

Behind the human couple is a rendition of Pioneer. This will give ET an idea of the general size of adult humans.

Beneath the human couple is a representation of our solar system. ET could use this to learn that Pioneer was constructed by beings on the third planet from the Sun and ejected from the solar system using a gravity assist from Jupiter, the fifth planet.

The mysterious spider-like construction in the left foreground of the plaque is an information-rich device. This pattern presents (in binary notation) the positions (from a terrestrial perspective) and periods of 14 radio-emitting pulsars—rapidly rotating remnants of exploded stars. Using this "spider" and the binary-translator elsewhere on the plaque, ET could pinpoint Earth's galactic location and, since pulsar emissions evolve with time, the approximate galactic era in which the Pioneer was launched.

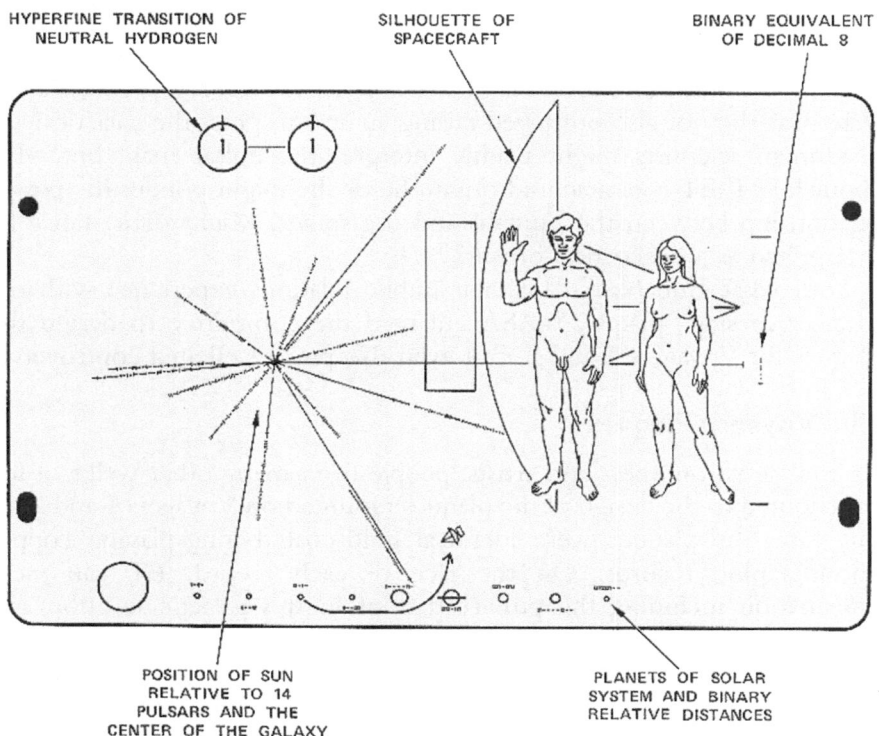

HYPERFINE TRANSITION OF
NEUTRAL HYDROGEN

SILHOUETTE OF
SPACECRAFT

BINARY EQUIVALENT
OF DECIMAL 8

POSITION OF SUN
RELATIVE TO 14
PULSARS AND THE
CENTER OF THE GALAXY

PLANETS OF SOLAR
SYSTEM AND BINARY
RELATIVE DISTANCES

FIGURE 14.1 Message plaque on Pioneers 10 and 11. (Courtesy NASA)

One would think that influential and educated humans would uniformly applaud this plaque as an appealing greeting card to our galactic brethren. But one would be wrong.

Look closely at the two human figures. Although the male is anatomically correct, the female has no vaginal cleft. This was airbrushed out prior to launch at the insistence of some American religious fundamentalists, who complained about sending pornography to the stars. This certainly says something about the American mind—nude females are pornographic, nude males are not!

When this controversy reached the public, the American left reacted with hilarity. One commentator suggested that all sexual references should be removed and that human reproduction should be alluded to with pictures of Santa Claus, the Tooth Fairy, and the Easter Bunny!

But even very thoughtful people realized that the best-designed interstellar message plaque would be somewhat ambiguous. Although humans might universally consider the male's upraised arm as a signifier of good will, how might extraterrestrials interpret it, especially those who had evolved from species equipped with poison darts launched from the armpit?

Some feminists also found fault with the female figure's passive stance. Why was she not also portrayed raising an arm to greet the galactics?

Human scientists might readily interpret the pulsar map, but what about ET? If ET is of arachnid origin, he or she might ponder the power relationship between the humans and the spider. Might their stance be interpreted as a declaration of war?

Somewhat embarrassed by their public relations experience with the Pioneer message plaque, NASA enlarged the committee to design the plaques for Voyagers 1 and 2. But even this process elicited controversy.

The Voyager Plaques

As well as astronomers and artists, people from many other walks of life contributed to the design of the plaques mounted on Voyagers 1 and 2. In this case, the plaques were identical gold-coated long-playing copper phonographic records. On the face of each record, ET can view information including the pulsar chart of Earth's galactic location and instructions for playing the record.

On each record, in digitized form, are more than 100 photographs representing aspects of our planet, environment, people, and civilization. Greetings from many human leaders (and one whale) in many languages can also be accessed, as can selections of human music.

But after all this effort, the Voyager plaques still elicited controversy. Why, for instance, was the only example of mid-twentieth-century western music a selection of Chuck Berry's work? And why was Beethoven's 5th Symphony selected rather than his 9th?

But NASA cannot fairly be faulted for the greatest controversy of all. No one could have foreseen that documents would later accuse the Secretary General of the United Nations during the mid-1970s, whose greetings were encoded on the Voyager plaques, of having had NAZI connections during the Second World War.

It is also worth noting that the cutting-edge 1970s audio technology used to create the Voyager records would soon become obsolete. Author Matloff remembers a cartoon showing the retrieval of Voyager by an ET species using digital audio-recording technology. Not knowing what to do with the gold recording, members of the starship crew use it in a game of zero-gravity Frisbee!

The 2006 launch of NASA's New Horizons mission to Pluto carries on it a new message to the stars—the ashes of Pluto's discoverer, Clyde Tombaugh. The ashes were donated to the mission by Tombaugh's family, and are sealed in a small aluminum canister on board the craft. Near the canister is the following message to some future generation of explorer, "Interned herein are remains of American Clyde W. Tombaugh, discoverer of Pluto and the solar system's 'third zone.' Adelle and Muron's boy, Patricia's husband, Annette and Alden's father, astronomer, teacher, punster, and friend: Clyde W. Tombaugh (1906–1997)." The mission was launched in the year of Tombaugh's 100th birthday.

A FUTURE MESSAGE PLAQUE POSSIBILITY: HOLOGRAPHY

Even though the interstellar message plaques currently in flight have been controversial, it is nearly certain that future extrasolar ventures will be equipped with such devices. Future message plaque designers may well decide to further broaden the base of participation, perhaps to include input from thousands of people. What is the best medium to contain such a comprehensive message, without exceeding the stringent mass limitations on the payload?

It should be a medium that could encompass creative input from thousands and it should be easily decipherable by extraterrestrials who intercept it. Future plaques should also be as low mass as possible, since mass budgets on near-term interstellar ventures will be extremely limited. Also, since a message plaque must survive for eons in interstellar space, it must be able to withstand the rigors of the space environment. And wouldn't it also be nice if the message plaque could serve a dual function such as assisting with spacecraft propulsion or steering?

Substantial progress toward creation of such a miracle message plaque began in the summer of 2000, when artist C Bangs curated an exhibition in conjunction with an International Academy of Astronautics symposium on deep-space exploration, in Aosta, Italy. About 35 visual artists contributed work to this "Messages from Earth" exhibition, which investigated visual concepts for future interstellar message plaques.

One of the symposium participants who viewed this exhibition was noted space scientist Robert Forward, who suggested to Bangs that white-light holography would be an excellent medium for future interstellar message plaques. Subsequent discussions between Forward and NASA managers resulted in a commission to Bangs to create a prototype holographic message plaque for the NASA Marshall Space Flight Center (MSFC).

It was soon apparent that creation of a hologram is labor intensive and not inexpensive. Very few artists will produce holograms in their studios, without the assistance of a skilled team of technologists.

It was first necessary to create some three-dimensional and two-dimensional images for incorporation into the holographic images. This aspect of the work was performed by C Bangs, with the assistance of computer artists David Wister Lamb and Lajos Szobozlai. The creation of the hologram itself was accomplished at the facilities of The Center for Holographic Arts, in Long Island City, New York. Prior to its delivery to NASA MSFC, the 40 by 50 centimeter hologram plate was framed by Simon Liu, Inc. in Brooklyn, New York.

All holograms are produced by the interference of two collimated light beams (helium–neon laser beams in this case). These monochromatic beams are usually produced by splitting the output of a single laser into two beams. One beam interacts with the object to be imaged. The second reference beam is then recombined with the first. A transparent photograph of the interference pattern is then used to produce the three-dimensional holographic image.

A traditional monochromatic hologram can only be viewed by inserting the transparent plate in the laser beam. A white-light hologram can be viewed in polychromatic, or natural, light.

More care must be taken with white-light holograms than with their monochromatic cousins. If a monochromatic transparent holographic plate is dropped and broken, each fragment can serve as a monochromatic hologram. If the same unhappy fate befalls your white-light hologram, all you have is a lot of broken glass.

The type of white-light hologram selected for the NASA prototype holographic message plaque was a Benton or rainbow hologram. Three-dimensional imagery is best incorporated in this variety of white-light hologram.

A non-holographic color image of the full Earth taken from an Apollo spacecraft in cis-lunar space serves as a backdrop in the prototype holographic message plaque. Holographic images include three-dimensional sculpted representations of male and female figures, two-dimensional line drawings of the same figures, a rendering of a hypothetical spacecraft trajectory and some of the relevant equations of solar-sail radiation-pressure acceleration.

View angle and illumination angle are critical in image visibility. It is generally necessary to shift view position to shift among images on the hologram. Under certain illumination conditions, two superimposed holographic images are visible.

Research revealed that rainbow holograms have an enormous information capability. According to Dr John Caulfield of Alabama A&M University, a well-designed hologram can contain hundreds of images, since a viewer angular shift of one degree or less is sufficient to shift between images. Since each image can be three-dimensional and contain many independent components, the work of many thousands of artists can be incorporated in a single message plaque.

As Caulfield pointed out, a space-qualified white-light hologram can be exposed on a photographic emulsion less than a micron in thickness. Thus, even a large message plaque need not add a great deal to the mass budget of a future extrasolar probe.

NASA propulsion engineers realized the possible utility of holographic films on solar sails. Since holographic film reflectance can be theoretically varied greatly by rotating a sail slightly from a reflecting surface's image to an absorbing surface's image, a holographic sail would allow thrust to be easily varied. Tests at the MSFC Space Environments Facility on commercial white-light holograms revealed that maximum hologram reflectance compares favorably with the reflectance of conventional sail films. Also, experiments revealed that holographic images are essentially immune to very large levels of simulated space radiation.

We see, therefore, that white-light holograms can be applied to both message plaques and propulsion. Settling the solar system need not be the sole province of engineers and scientists—artists will likely be right beside them! Creation of this prototype message plaque was a very interesting exercise on the boundaries of art and science.

FURTHER READING

The best popular description of the Pioneer plaques is included in Carl Sagan's *Cosmic Connection* (Doubleday, Garden City, New York, 1973). To learn more about the Voyager plaques, consult C. Sagan, F.D. Drake, A. Druyan, T. Ferris, J. Lomberg and L.S. Sagan, *Murmurs of Earth* (Random House, New York, 1978).

Various aspects of the prototype holographic message plaque created for the NASA Marshall Space Flight Center and subsequent research on holographic applications to spacecraft propulsion are reviewed in G.L. Matloff's *Deep Space Probes*, 2nd edn (Springer–Praxis, Chichester, UK, 2005). Another source is G.L. Matloff, G. Vulpetti, C Bangs and R. Haggerty, *The Interstellar Probe (ISP): Pre-Perihelion Trajectories and Application of Holography*. NASA/CR-2002-211730 (NASA Marshall Space Flight Center, Huntsville, AL, 2002).

15

SPACE BEANSTALKS

Higher still and higher
From the earth thou springest
Like a cloud of fire;
The blue deep thou wingest

Percy Bysshe Shelley, from *To a Skylark*

IMAGINE catching a plane to a Pacific island located on the equator, walking from the airplane to an elevator that goes straight up—into space—taking you from the ground to any altitude up to or below geosynchronous orbit (that is, approximately 35,786 kilometers—the altitude at which an Earth-orbiting satellite travels with a velocity that maintains it over a fixed point as the Earth rotates beneath it). If you were to travel the full 35,786 kilometers in the elevator, you too would be in orbit around the Earth—and paying just the cost of the electricity operating the elevator to get you there. And this would all be as effortless on your part, and as reliable, as the flight of a skylark.

It is not a new idea. In fact, a visionary Russian scientist named Konstantin Tsiolkovsky first proposed it over a century ago. He claims to have thought of the idea while looking at the Eiffel Tower. He imagined a cable running from the Earth to space, at the end of which would be a way station for space travelers. Is such a tower possible? If it were, would it be the "best" way to get into space? What would be the implications for our exploration of space? These questions are still being debated today. There is no doubt that it would be a cost-efficient, low-pollution method of space access, so it merits some discussion.

If we assume that a space elevator can be constructed, what would it have to be made of in order to not collapse under its own weight or, alternatively, be pulled apart while it is under construction?

Even if we could develop the super-strong, very low-mass material out of which a ground-to-geosynchronous space elevator could be constructed, there are a number of very major obstacles, one of which is what to do about orbital debris.

Because the top of a space elevator is in geosynchronous orbit, the entire structure will seem to hover over one point on Earth's surface. Satellites in low Earth orbit will whiz past the tower's structure at about 8 kilometers per second. If a small piece of orbital debris struck the elevator, the structure would be severed. The entire 35,000 kilometer length of the severed space elevator might then wind itself around Earth's equator several times, with unfortunate consequences for dwellers of equatorial countries such as Ecuador. So we might initially consider instead some less ambitious applications of long, thin cables in space.

ELECTRODYNAMIC TETHERS: TAPPING A PLANET'S MAGNETIC FIELD FOR POWER AND PROPULSION

An electrodynamic tether, which is basically a long electrically conducting wire, can propel a spacecraft by virtue of the force a magnetic field exerts on the wire when it is carrying an electrical current. This phenomenon was first explained scientifically by Ampere, one of the pioneers in the study of electromagnetism, around 180 years ago. The physics describing the nature of this force, which acts on any charged particle moving through a magnetic field (including the electrons moving in a current-carrying wire), was described by Hendrik Lorentz in 1895 in an equation that now bears his name. The force acts in a direction perpendicular to both the direction of current flow and the direction of the magnetic field (Figure 15.1). Electric motors make use of this force: a wire loop in a magnetic field is made to rotate by the torque the Lorentz force exerts on it due to an alternating current in the loop interacting with the field

FIGURE 15.1 *The Lorentz force (F) on a charged particle (P) is perpendicular to the magnetic field and the particle velocity V (the direction of current flow). The magnetic field direction is down, into the page.*

produced by a magnet. The motion of the loop is transmitted to a shaft, thus producing a motor. Michael Faraday demonstrated this first electric motor about 1821.

Given that the fundamental physics was worked out almost two centuries ago, the working principle of an electrodynamic tether propulsion system is not new, but its application to space propulsion may change the way we plan space missions. In essence, a tether thruster is just a clever way of getting an electrical current to flow in a long orbiting wire (the tether) so that the Earth's magnetic field will accelerate the wire and, consequently, the payload attached to the wire. The direction of current flow in the tether, either toward or away from the Earth along the local vertical, determines whether the magnetic force will raise or lower the orbit. The key point is that all of this is done without the need for fuel. An electrodynamic tether provides propulsion by interacting with the natural space environment and does not require any resupply! They have even been tested in space.

To understand how they work, a short description of the space environment in which they operate is in order. First of all, electrodynamic tethers work best when they are in the presence of space plasma. Plasma is nothing more than a collection of freely moving ions and electrons. In a system without external energy coming in, these ions (which are positively charged) and electrons (which carry a negative charge) would attract one another and produce electrically neutral atoms without a net charge. When energy is freely available—typically from the Sun in the form of ultraviolet light—electrons are stripped from atoms, creating the plasma and sustaining it. The residual atmosphere in low Earth orbit contains plasma. Another natural environmental factor required for electrodynamic tether propulsion is a magnetic field. The Earth is surrounded by a fairly strong magnetic field that is thought to be produced by our planet's molten iron core,. The strength of the field decreases with distance, making it fairly strong in low Earth orbit and weak at geostationary orbital altitudes.

Electrodynamic tethers were successfully demonstrated in space by flights of the Plasma Motor Generator (PMG) in 1993[1] and the Tethered Satellite Systems (TSS-1 and TSS-1R) in 1992 and 1996.[2,3] All three missions deployed long conducting tethers from orbiting spacecraft and successfully generated a current, though none used the current to provide measurable propulsion. Readers with a background or knowledge of electricity and magnetism will immediately question how this system can produce net thrust since a loop is required for current to flow. Attaching a wire to only one terminal on your car battery does not produce a current flow—the wire(s) must be connect to both terminals or one terminal and the ground. In such a loop, the force on the tether resulting from the current flowing through it in one direction would be exactly canceled by the current flowing back through the wire loop in the other direction (which "closes" the circuit) producing no net thrust! The answer is that, in space, the tether forms only one half of the loop; the space plasma in the ionosphere forms the other. The tethered system extracts electrons from the plasma at one end (upper or lower, depending upon the deployment direction and intended thrust motion) and carries them through the tether to the other end, where they are returned to the plasma. Currents in the plasma then complete the circuit. The net force caused by a magnetic field acting on a current-carrying closed loop of wire (i.e. a normal circuit) would be zero, since the force on one length of the wire would be canceled by that on the other through which the current was flowing in the opposite direction. However, since there is no mechanical attachment of the tethered system to the plasma, magnetic forces on the plasma currents do not affect the tether motion. In effect, we have a length of wire with a unidirectional current flowing in it, and this wire is accelerated by Earth's magnetic field (Figure 15.2).

Imagine a spacecraft in low Earth orbit, traveling at 8 kilometers per second immediately after being launched into space by a rocket. A spring

[1] McCoy, J.E. et al., "Plasma Motor Generator (PMG) Flight Experiment Results," *4th International Conference on Tethers in Space, Smithsonian Institution*, Washington, DC, 10–14 April, 1995.
[2] Strim, B., Pasta, M. and Allais, E., "TSS-1 vs. TSS-1R," *4th International Conference on Tethers in Space, Smithsonian Institution*, Washington, DC, 10–14 April, 1995.
[3] Raitt, W.J. et al., "The NASA/ASI TSS-1 Mission: Summary of Results and Reflight Plans," *4th International Conference on Tethers in Space, Smithsonian Institution*, Washington, DC, 10–14 April, 1995.

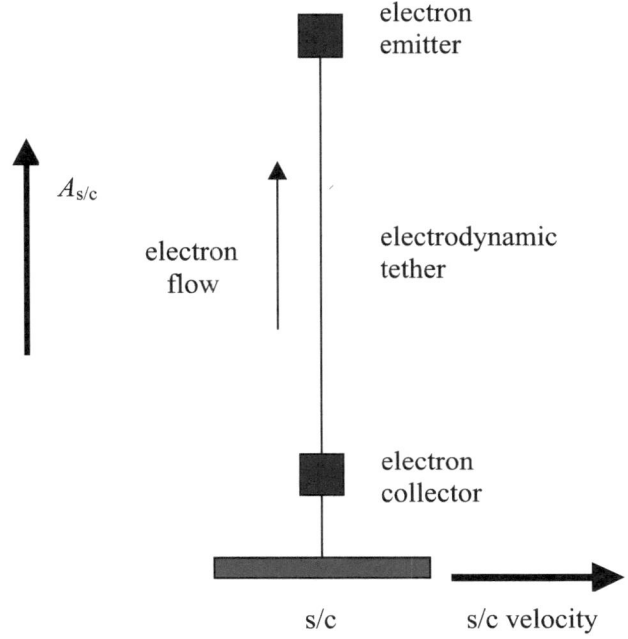

$A_{s/c}$

electron
flow

electron
emitter

electrodynamic
tether

electron
collector

s/c

s/c velocity

Earth

FIGURE 15.2 An electrodynamic tether in low Earth orbit, with a unidirectional current flow. The spacecraft (s/c) is attached to the tether. The spacecraft's acceleration ($A_{s/c}$) is caused by the interaction with Earth's magnetic field.

ejects upward a small payload attached to the spacecraft by a conducting tether. The tether deploys upward from the spacecraft by the forces exerted due to the difference in gravitational attraction between the payload and the spacecraft (technically known as the "gravity gradient force"). This occurs because the force of gravity decreases with distance. And the distance created initially by the spring ejection is sufficient to produce a significantly different gravitational force acting on the spacecraft versus the payload—effectively pulling the tether from a spool to virtually any length desired. Tethers as long as 20 kilometers in length have been deployed this way in space.

But what is the source of the voltage required to drive the current? The voltage across a vertically deployed conducting tether, which results just from its orbital motion through Earth's magnetic field, is positive at the top and negative at the bottom. This is once again due to the Lorentz force acting on the electrons in the tether. The magnetic force on the tether wire has a component opposite to the direction of motion, and therefore slows down the system, lowering its altitude, leading eventually to reentry. In this "generator" mode of operation the Lorentz force serves both to drive the current and then act on the current to decelerate the system. The TSS and PMG missions described above demonstrated this in space, but, as stated previously, no measurements were made to quantify the relatively small orbital changes. And that was a shame!

In addition to not requiring any fuel, no onboard power source is required to drive the electrical current flow in either the orbit-raising or orbit-lowering modes. Again, the environment provided by nature gives us all we need—sunlight. To raise the orbit, sunlight falling on a solar array can be converted to the electrical energy required to drive the tether current in the opposite direction. To lower the orbit, the power comes from the orbital energy of the spacecraft (supplied by the Earth-to-orbit launcher when it placed the system into orbit).

Having established that an electrodynamic tether system can provide essentially "free" propulsion around any planet with both a plasma and a magnetic field, what might it be used for? The short answer is "lots of things!" From extending the orbital lifetime of large space structures like the International Space Station, to providing a lightweight deorbit system for use at the end of a space mission (so as to not create more space junk), they can be used by virtually any spacecraft in low Earth orbit.

The International Space Station, or ISS, experiences significant aerodynamic drag in its approximately 400-kilometer altitude orbit. In this low Earth orbit, the vacuum is not perfect. At space station altitudes, the atmospheric pressure is very low, but it is not zero—contrary to popular misperception, Earth orbit is not a perfect vacuum! When the ISS, having a very large surface area, passes through this tenuous gas traveling at 8 kilometers per second, it experiences a continuous drag force of approximately 1 newton. The drag is caused by the collision of the space station with the residual atmosphere. It is no different from the drag a car experiences when it is moving through the air on the freeway—though the magnitude of the drag force in space is significantly less because there is so little atmosphere at these altitudes. It is a very small force, approximately the same force that a piece of notebook paper exerts on the palm of your hand. But it is a constant 1 newton force, causing a slow but steady braking force on the station. As a result, the space station's altitude slowly decays until a reboost maneuver is performed. Without reboost, the ISS would sink ever lower until it finally burned up in the atmosphere. Reboosting something as massive as the space station requires a propulsion system and, conventionally, fuel—lots of fuel.

Outfitting the space station with a tether propulsion system would significantly reduce or eliminate its dependence on Earth-launched propellant resupply, which is needed to accomplish the required reboost.[4] The ISS can supply its own electrical power using solar arrays but not its own propellant. A 7-kilometer tether and 6 kilowatts of electrical power could reduce the number of fuel resupply flights to the ISS, potentially saving $1 billion over a decade of operation! Even if the planned frequency of resupply flights to the ISS is maintained, putting a tether propulsion system on board would then provide the option to trade power for increased payload capacity. Resupply vehicles might then deliver useful cargo like scientific instruments, replacement parts, and crew supplies rather than fuel.

Another critical application of tether propulsion might be for use during the assembly of large space vehicles going to the Moon or Mars. Many of the options being considered for such exploration involve putting together large vehicles from many smaller pieces launched from Earth. A tether reboost system would allow these pieces to remain in orbit indefinitely while the remaining pieces are launched from Earth and thus provide mission flexibility if a launch delay is encountered. And no launch costs would be incurred for the massive amounts of fuel that would otherwise be required.

A tether might also be used on an orbital tug to move payloads in LEO after launch. The tug would rendezvous with the payload and launch vehicle, dock with the payload, and maneuver it to a new orbital altitude or inclination within LEO *without the use of boost propellant*. The tug could then lower its own orbit to rendezvous with the next payload and repeat the process. Such a system could perform multiple orbital-maneuvering assignments without resupply, substantially lowering costs.[5] Current technology requires that a new upper stage be built, used, and thrown away for each of these mission phases.

Tether propulsion would also be useful for one of the most propulsively intense orbital operations—changing a spacecraft's orbital inclination.

---

[4] Johnson, L. and Herrmann, M., *International Space Station Electrodynamic Tether Reboost Study*. NASA/TM-1998-208538.

[5] Johnson, L., "The Tether Solution," *IEEE Spectrum*, July 2000.

When a rocket is launched from the ground, some of the Earth's rotational energy is carried with it, lowering the amount of energy that must be added to achieve orbital velocity. This is why spacecraft are almost always launched eastward, and not westward, where the rocket would be acting *against* the Earth's rotation. The launch point also determines the inclination, or tilt relative to the ground, of the spacecraft's orbit as it moves around the Earth. Recall that the Earth is tilted on its axis by approximately 24 degrees. (It is this axial tilt that provides our seasons.) As a spacecraft orbits, the tilted Earth rotates beneath it, circumscribing an orbital trace that appears to move up and then down across the face of the Earth. The higher the latitude of the launch, the more northerly the trace appears. This is the orbital inclination of the spacecraft, and to change this inclination in space after launch requires a large amount of propulsive thrust, hence propellant. Given that most US launches occur in Florida, which has a 28.5-degree latitude, the easiest attainable orbit has a 28.5-degree inclination. A spacecraft equipped with an electrodynamic tether could selectively flow current through the wire to maximize the thrust out of the orbital plane, resulting in an inclination change. And no fuel would be required!

Electrodynamic tether propulsion does, however, have limits. It can only be used where the magnetic field is strong and the plasma current densities are rather large. For Earth, this limits their use to between approximately 400 and 2,000 kilometers. Below 400 kilometers, the atmosphere is too dense and the atmospheric drag on the tether exceeds its ability to produce thrust. Above 2,000 kilometers the plasma density gets so low that there are simply not enough electrons to collect and produce the required current.

TETHERS FOR PROPULSION AND POWER AT JUPITER

Following NASA's successful Galileo mission to Jupiter, there is considerable interest in going back to study the planet and its moon, Europa. Galileo, which was launched in 1989, arrived at Jupiter in 1995. The spacecraft, which was about the size of a school bus, remained in orbit, taking breathtaking pictures and gathering an immense volume of scientific data until 2003, when the command was given to end the mission by crashing the spacecraft into Jupiter's atmosphere. The Galileo mission revolutionized our understanding of Jupiter and its satellites, including the mapping of its large and very strong magnetic field and ionosphere. This mapping will be vital for future missions to Jupiter as it holds the key to tapping its immense energy for producing both power and propulsion.

The Galileo spacecraft, like many spacecraft sent to the outer planets, was mainly a propulsion system. In fact, more than 50% of the spacecraft's weight at launch was devoted to simply "getting there" and stopping upon arrival. In order to escape the Earth's gravity and gain enough speed to travel the 4 AU distance to Jupiter, large amounts of propellant are required. Not only must a spacecraft attain a very fast velocity in order to keep the trip time reasonable (it took the Galileo spacecraft 6 years to make the journey to Jupiter), but the spacecraft must use still more propulsion to enable it to slow down enough to go into orbit around the planet. Without slowing down, the spacecraft would simply fly right by the planet on its way out . . . to somewhere else. As fantastic as the mission was, wouldn't it have been even better if less propellant had been required, allowing the inclusion of more science instruments for a yet greater scientific return? In addition, extended operation in the Jovian system, or around any planet, typically requires the use of propellant for orbital maneuvering. This is especially important in a system as interesting and as complex as Jupiter. The need to perform such maneuvers leads to yet more of the very limited spacecraft mass at launch being allocated to propulsion or, with less propellant, a limited lifetime on orbit—stopping the mission as soon as the fuel runs out.

In addition to the requirements of the propulsion system, the mission was limited by power. Due to minimal sunlight falling on any reasonably sized solar array so far from the Sun (approximately 483 million miles), radioisotope thermoelectric generators (RTGs) were used for electrical power by Galileo and in all past deep space missions. The perceived risk of releasing plutonium into the terrestrial environment during either launch or on an Earth gravity-assist "flyby" complicates the use of RTGs on future missions. The possibility of using solar panels for electrical power generation has improved in recent years due to significant advances in solar array materials; however, the high levels of radiation in the Jovian system will likely limit their lifetime. Solar array materials and ionizing radiation do not mix well.

For these reasons, and because of the strong Jovian magnetic field, electromagnetic tethers are being studied for producing power and propulsion on spacecraft in orbit around Jupiter.[6] Studies show that 1

---

[6] Gallagher, D.L., Johnson, L., Moore, J. and Bagenal, F., *Electrodynamic Tether Propulsion and Power Generation at Jupiter.* NASA/TP-1998-208475.

megawatt of power can be theoretically generated by a 10-kilometer tether in near Jovian space—a tether similar in length to those that have been used in Earth orbit. Specifically, such a tether operating near Jupiter would experience induced voltages greater than 50,000 volts, currents in excess of 20 amperes, generate approximately 1 megawatt of power and experience more than 50 newtons of thrust! Needless to say, this would pose significant engineering challenges, lest the engineers end up with the brightest fuse burning anywhere in the solar system, ending any mission before it has a chance to begin!

In addition to the very large power generated, other technical challenges will face those using tethers to harvest energy from the Jovian environment. One such difficulty is the gravity gradient. The slight difference in gravitational pull across the length of the tether is what keeps it taut while thrusting or generating power. Without it, the tether would flop around like a noodle in the breeze and not be of particular use to anyone. While Jupiter is the most massive planet in our solar system, it is also the largest. That means that its gravity gradient-induced force on the tether, or the difference in the gravitation attraction from one end of the tether to the other, is relatively small. The solution might be to spin the spacecraft to enable centripetal force to keep the tether taut. That, of course, complicates just about everything—but with the promise of unlimited power and propulsion, it is a problem that is certainly worth solving.

Unfortunately, electrodynamic tethers cannot be used in interplanetary space or for providing the propulsion necessary to get from one planet to the next. The reason is simple: there is neither an appreciable magnetic field nor a significant plasma with which to interact in deep space. Tethers such as these are confined to operation near or around planets that have both magnetic fields and space plasma.

There is a way, however, to take advantage of two propellantless propulsion systems working in tandem to make the journey from Earth to Jupiter, stopping there to do science, while getting all the propulsion and power needed from the abundant resources of space itself. This might be achieved by the marriage of electrodynamic tethers and solar sails. A large solar sail, with electrodynamic tethers embedded in it like the spokes on a bicycle wheel, would use the copious sunlight available in the inner solar system to get up to speed (see Chapter 13 for a discussion of solar sails) and the interaction of the tethers embedded in the sail with the Jovian magnetic field and space plasma to slow the spacecraft and allow it to be captured into orbit. The sail would spin, providing the centripetal acceleration needed to keep the thrusting tethers taut, as the system is captured into orbit (Figure 15.3).

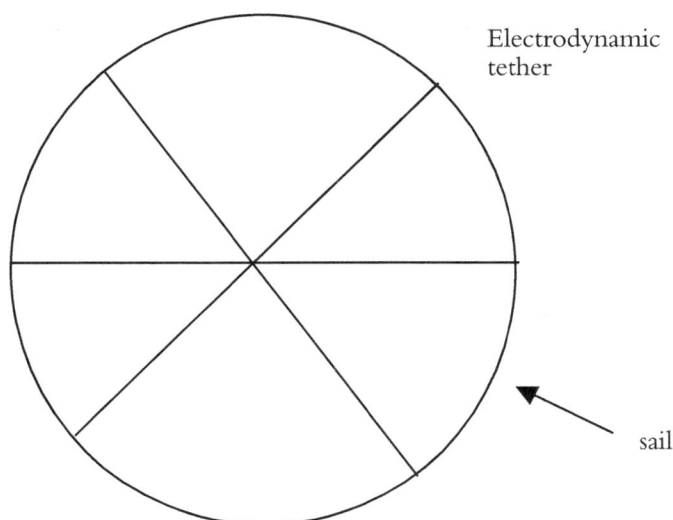

FIGURE 15.3 An electrodynamic tether (ET) integrated with a solar photon sail.

MOMENTUM EXCHANGE ELECTRODYNAMIC REBOOST (MXER) TETHERS

Imagine a propulsion system that could send spacecraft from low Earth orbit, or even from high within the Earth's atmosphere, to the Moon, Mars—or beyond—without using propellant. Such a system would eliminate the need to carry massive amounts of fuel from the ground into space just to get enough speed to make the journey and to slow down and stop at the destination. Then, once the mission is complete or the product is ready to return to the Earth, a variation of the same propulsion system would pick up the payload and send it home. All this is possible using a related, but different type of tether system called a Momentum Exchange Electrodynamic Reboost (MXER). Once implemented, the MXER may become the railroad of space.

The physics is simple. Recall your childhood yoyo and how you probably played with it, learning to do tricks until one day you got a little bored and decided you were going to play "David and Goliath"—using the yoyo as David's slingshot. Spinning it rapidly over your head you let go of the string and watched the yoyo become a projectile that hopefully didn't hit your kid brother or sister in the eye! This is the very same principle by which a MXER tether can be used to send a spacecraft payload on its way to anywhere in the solar system. A large, rotating tether system in space could be used to "throw" a spacecraft out of the Earth's gravity and on its way to the Moon, Mars, or virtually any other destination in the solar system—and then be used again for the next spacecraft.[7]

To explain how the overall "system" works requires first a careful explanation of each of its components. The key elements are: (1) the non-conducting portion of the rotating tether; (2) the electrodynamic portion of the tether; and (3) the power system. The first component is the bulk of the rotating tether system. A tether is nothing more than a cable connecting two or more bodies in space. In this case it must be relatively long (approximately 100 kilometers in length) and strong enough to not break during operation. Fortunately, modern materials are available that meet these requirements. To avoid being cut by micrometeors or debris in Earth orbit, the tether must either be braided into a rather substantial rope, larger in diameter than any of the particles or objects likely to hit it during operation, or designed such that any cut portion of the tether results in the loads being redistributed to other sections, keeping the system intact (see Figure 15.4). The latter approach, called a "Hoytether" is named after its inventor, Rob Hoyt. Rob's design looks like a ladder. If one piece is destroyed by a micrometeor, then the other parts of the ladder will carry the load and allow the overall tether to survive. Make no mistake, anything hit by a piece of debris traveling at 29,000 kilometers per hour will be damaged. The width of the braided tether will do nothing to help the portion that is hit to "survive" the impact, rather the width of the tether is large so that any hits will only "nick" the tether, leaving it mostly intact. This is because most of the micrometeors and debris are fairly small compared to the width of the more distributed Hoytether.

The MXER tether does not work by magic. Newton's laws apply, and once a payload is thrown it gains energy and the MXER tether station loses energy. The payload's energy increase is easily seen—it speeds up and is moving quickly away into interplanetary space. Where did the energy come from? The spacecraft's increase in kinetic energy came from a decrease in the MXER tether station's kinetic energy, hence its orbital altitude. In order for the station to throw another payload, it must regain energy and boost back to where it was before its last use. This is accomplished by the use of the electrodynamic, or conducting, tether embedded within the length of the overall MXER tether. This relatively

[7] Sorensen, K., *Conceptual Design and Analysis of an MXER Tether Boost Station*. AIAA Paper 2001-3915.

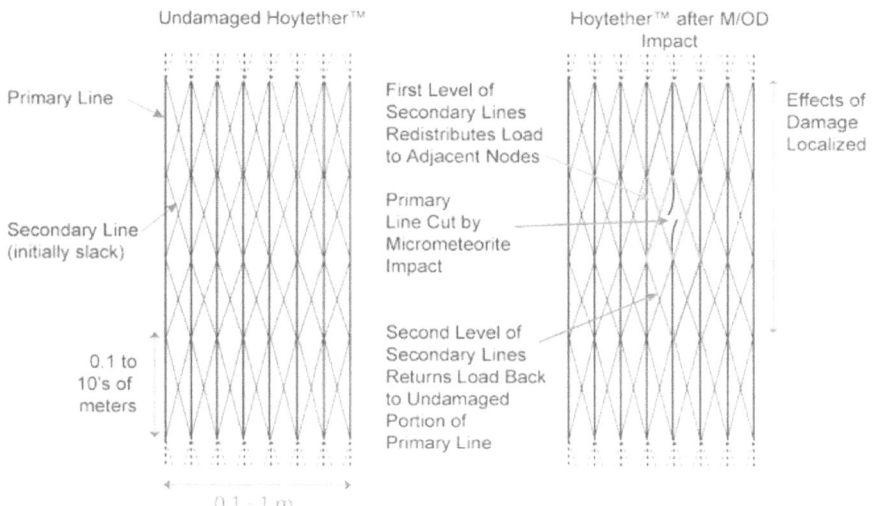

FIGURE 15.4 The Hoytether: a multi-strand failure-resistant tether structure. The acronym "M/OD" stands for "meteroid/orbital debris". (Courtesy Dr Robert Hoyt, CEO, Tethers Unlimited)

small conducting section of tether, if power is provided, can push against the terrestrial magnetic field and reboost the MXER tether station to its operational altitude. Once the reboost maneuver is complete, MXER tether station will be ready and "recharged" to send another spacecraft on its way.

Where does the power required to reboost the system come from? In Earth orbit, solar arrays are capable of providing more than enough power to run the station and reboost it, as required. Ultimately, the MXER tether station will rendezvous with a spacecraft, capture it, spin it up, release it and then recover lost altitude and energy without the expenditure of propellant. The entire system operates on renewable resources, potentially allowing it to run autonomously for years. The cycle of operation of a MXER boost station is shown in Figure 15.5.

In addition to operating MXER tether stations in Earth orbit, one can be placed in low lunar orbit. The station orbiting the Moon cannot operate as efficiently as the one near Earth due to the Moon's lack of both an appreciable magnetic field and ionosphere. The lack of these resources makes it impossible for a lunar tether to reboost using electrodynamic tether forces. Replacing the conducting tether component of the system with a highly efficient electric propulsion system (for reboost) would still make it an attractive option for sending spacecraft and payloads on their

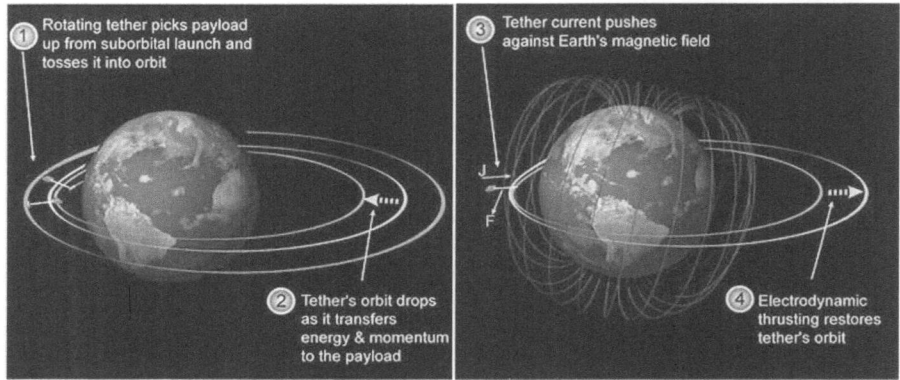

FIGURE 15.5 The Orbit of a MXER tether boost station before and after operation. (Courtesy Dr Robert Hoyt, CEO, Tethers Unlimited).

return journey to Earth. This is especially true when you consider that the Moon has no atmosphere and therefore there is no reason that a long tether could not come very close to its surface. A tether dipping low into the Earth's atmosphere would experience incredible drag forces and be subjected to severe heating from its high-speed passage through it— probably resulting in the tether melting or otherwise breaking. This would not occur on the Moon. In fact, if the rotational velocity of the tether is controlled sufficiently, the tip of the tether could be made to dip down and touch the surface of the Moon—having for a moment near-zero relative motion compared to that point on the surface—thus potentially picking up any payload directly from the surface of the Moon and carrying it into space. Once there, it could be tossed toward its Earthly counterpart. Instead of relying on chemical rockets to get from the surface of the Moon to space, as was done in Apollo, a tether can simply come down from the sky and provide a relatively free ride home.

Consider how this might look to our lunar astronauts. After placing the cargo for return to Earth in a designated drop zone, they move back and watch as a rope descends from space, swinging across the horizon, appearing to slow down at the drop zone while the cargo is clamped to it. It then slowly accelerates along its arc back into space—carrying the cargo with it. The tether doesn't really slow down, of course, but it appears to do so because at that moment, both it and the Moon are rotating and have components of their velocities in the same direction.

There is another way that either or both of the MXER tether stations can gain energy and regain operational status following a "throw." If they "catch" an incoming payload, the kinetic energy of motion from the captured spacecraft would be conserved, resulting in the energy of the

MXER tether station increasing. The energy gained during a catch would reboost the station, minimizing or eliminating the need for the electrodynamic tether or electric propulsion system to do so.

MXER tether systems are not without their problems; fortunately few of them are really technical. For example, to maximize the utility of a MXER tether system, it would need to be placed in orbit above the equator. The United States has a very limited ability to launch spacecraft into equatorial orbit given that most of our launches occur in Florida, giving them an orbital inclination of 28.5 degrees. Either much time and energy would be wasted changing the orbital inclination of launched payloads in order for them to reach the MXER station, or a new launch facility would have to be constructed on the equator. Both of these options are expensive.

Nonetheless, one can easily imagine a network of such stations transferring payloads between the Earth and the Moon—efficiently and without the consumption of much propellant.[8] Compare this with our current capabilities to transfer spaceships between the Earth and Moon and you will quickly see its advantages. The entire mass of the Saturn V rocket was required to carry enough propellant to send the tiny Apollo spacecraft containing three people to the Moon and back again.

Never again need we be so wasteful when Mother Nature is ready to provide us with all the energy we require—she just waits for us to be clever enough to use the resources she has provided. The next step would be to place a MXER tether station in Mars' orbit—emplacing the next stop on our space railroad.[9]

FURTHER READING

A number of popular treatments of tethers and related devices have been published. Consult, for example, Chapter 2 of Giancarlo Genta's and Michael Rycroft's *Space: The Final Frontier?* (Cambridge University Press, New York, 2003).

The Johnson/Matloff tether/sail combination is further described in US Patent No. 6,497,355 B1. For further information on the Hoytether, consult the website: http://www.tethers.com/Hoytether.html

---

[8] Forward, R.L., "Tether Transport from LEO to the Lunar Surface," *AIAA-91-2322, AIAA/ASME/SAE/ASEE 27th Joint Propulsion Conference*, Sacramento, CA, 24–27 July 1991.

[9] Forward, R.L. and Nordley, G., "Mars–Earth Rapid Interplanetary Tether Transport (MERITT) System: Initial Feasibility Analysis," *AIAA-99-2151, AIAA/ASME/SAE/ ASEE 35th Joint Propulsion Conference*, Los Angeles, CA, 20–24 June 1999.

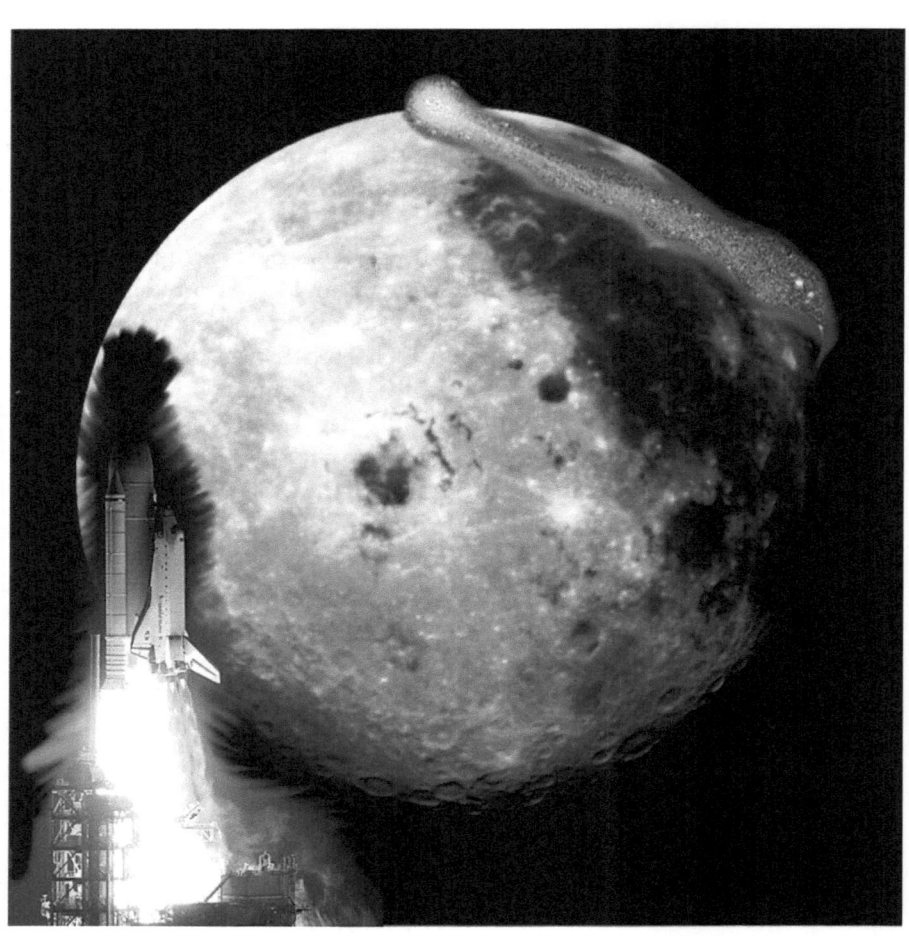

16

CHEMICAL PROPULSION FOR SPACE EXPLORATION:

GOOD FOR YESTERDAY, TODAY AND TOMORROW

Is there so small a range in the present strength of manhood,
that the high imagination cannot freely fly as she was wont of old?
Prepare her steeds, paw up against the light,
And do strange deeds upon the clouds?

John Keats, from *Sleep and Poetry*

Chapter 16

C HEMICAL propulsion has been the workhorse of space exploration since its beginning. Chemical propulsion is, as its name implies, all about chemistry and chemical reactions. Combine certain chemicals and chemical bonds form, releasing energy. It is this energy that is harnessed to provide a rocket's thrust. Given that chemical rockets are fairly simple, provide high thrust, and are mostly reliable, they will likely be the mainstay propulsion system for the foreseeable future. However, unless we change the way we use them, their implementation will become more and more of a liability, ... which requires some explanation.

Today, when we begin a mission to anywhere in the solar system, whether it be to low Earth orbit (LEO) or a flyby of Pluto, the spaceship is launched with a throwaway chemical rocket and sent on its way carrying all the fuel it needs—fuel produced on the Earth and packaged with the spacecraft, its launch vehicle, or upper stage. In theory, you can add more fuel or additional stages to increase the total propulsion provided to the spacecraft. This is a method with increasingly diminished returns, thanks to the Rocket Equation (see Chapter 3).

Basically, the Rocket Equation is nothing more than Newton's law applied to rockets: in order to produce an acceleration (*a*) of a mass (*m*), one needs to apply a specific force (*F=ma*). Simple; except that to produce the acceleration with a rocket involves a changing mass—the fuel load is part of the total mass being accelerated and is decreasing with burn time. Even worse, the addition of fuel to produce the force that provides the needed acceleration drives up the total mass, requiring you to add yet more fuel—and a circular, vicious cycle begins. More force required to accelerate means that you need more fuel, which means that you need more force, which requires yet more fuel. This continues until you reach the point when any fuel added is needed just to move its own mass—making is senseless to add it in the first place.

Early rocket pioneers worked around this problem by a process called *staging*. We take advantage of staging in virtually every mission launched today. If you add a stage, which includes not only the fuel but also the structure required to contain it, and then throw away the stage after it is used, the result is that you have less overall mass to accelerate and can reduce the overall mass (hence fuel) required to achieve the mission. This will help, but unless you can develop stages that have zero mass (including the fuel), the Rocket Equation will still lead to a point of diminishing returns.

So far, we've been discussing the physics and engineering limits imposed by the Rocket Equation. If we look at cost, which is currently hovering around $7,000–10,000 per pound of payload launched into space, we reach that point of diminishing returns by another route—our ability to pay for all the required fuel!

Clearly, if we are going to continue using chemical propulsion, we need to find a way by which it is NOT necessary to bring it all with us from the launch pad. In addition, many of today's rocket fuels are quite toxic, which greatly complicates the handling of the spacecraft and its propulsion system before launch and is a potential health concern to people and wildlife near the launch site. Monomethyl hydrazine, a commonly used onboard propellant (which is used on board the space shuttle for orbital maneuvering), is highly toxic and imposes significant additional costs on spacecraft manufacturers just due to the added complexity caused by its need for safe handling.

Ideally, we should use a chemical propulsion system that can be replenished in space, safely and affordably. Fortunately, nature may provide us with several options for doing this very thing. The list of available options depends strongly upon where we are in the solar system.

THE MOON: A REFUELING DEPOT FOR DEEP-SPACE EXPLORATION

The space shuttle main engines, among many other rockets, use the very simple chemical reaction between hydrogen and oxygen to produce thrust. When launched, the shuttle's main engines burn 530,000 liters ($\sim$140,000 gallons) of liquid oxygen and 1,400,000 liters ($\sim$380,000 gallons) of liquid hydrogen. This, combined with the orbiter's solid rocket motors, produces enough thrust force to lift the shuttle's 2-million-kilogram (4.4-million-pound) weight from the surface of the Earth to low Earth orbit. These two simple chemical ingredients, hydrogen and oxygen, may be available on the Moon for our use.

Lunar rock contains oxygen in abundance. Using concentrated sunlight, provided by solar concentrators like those described in Chapter 12 for solar-thermal rockets, the rocks could be superheated to the point where the oxygen is released and collected. This would be the easy part. Oxygen-bearing rock appears to be available anywhere on the Moon. Finding hydrogen on the moon will be much more difficult.

Hydrogen, with an atomic number of 1, is the lightest of all elements. Not much energy is required to get hydrogen atoms moving at speeds that exceed the lunar escape velocity (the speed at which an object will escape from the gravity of a planet or moon, never to return). Over the billions of years since its creation, any atmospheric hydrogen the Moon might once have had has long gone. On the Earth, in addition to atmospheric hydrogen, there is water. One water molecule is composed of two hydrogen atoms and one oxygen atom. Our vast oceans and lakes provide a huge supply of hydrogen to meet our needs. Until fairly recently, the Moon was considered the ultimate desert—totally bereft of water. Then the Lunar Prospector Mission provided tantalizing hints that water ice might be preserved deep in craters near the lunar poles. Lots of ice, perhaps as much as 300 million tons ...

If there is water on the Moon, then its hydrogen component can be easily obtained by passing an electrical current through the water in a process known as electrolysis, liberating both the hydrogen and oxygen. The process is not complicated and is used regularly in earthly chemical plants and science classrooms.

Once the two gases are collected, the ingredients are in place for fueling chemical rocket engines to continue our journey.

MARS: THE RED PLANET MAY PROVIDE FUEL FOR ROUND TRIP TRAVEL TO EARTH

The Martian atmosphere is composed of approximately 95% carbon dioxide, 2.7% nitrogen, less than 0.13% oxygen, and other trace gases. With the addition of hydrogen, all the ingredients needed to produce methane are in place. The methane/oxygen reaction works very well for rocket propulsion, producing only carbon dioxide and water as byproducts—no messy, toxic byproducts result.

The scenario might look something like this. A spacecraft descends to the Martian surface, does whatever it is designed to do (it can be either robotic or human-tended, the premise is unchanged), and is ready to return to space. Rather than load onto the spacecraft at Earth all the propellant needed to launch from the Martian surface back into space— lugging it on the voyage to Mars and paying a heavy (literally) price to land it on the surface—we launch the craft with its Mars ascent fuel tanks *empty*. We then land the spacecraft near some sort of tanking station on the surface that has been producing methane rocket fuel from *in situ* Martian resources for this very purpose. The methane is transferred to the ascent stage, the rocket is lit, and the payload returns to space.

To make this happen, a miniature chemical plant, using well-understood chemical processes, will be placed on the surface to produce rocket fuel. In order to be economical, the plant should be able to support one large human-scale mission or multiple robotic scale missions. The first ingredient, methane, is extracted from the carbon dioxide in the atmosphere by reacting it with hydrogen, producing water in the reaction. The hydrogen comes from one of two places: (1) since it is of low molecular weight, the spacecraft brings it along—which partially defeats the objective of weaning ourselves from the home planet, or (2) we get it from local Martian water. Recent observations of Mars are causing many to believe that large reservoirs of subsurface water might exist there. If it does, then the hydrogen component of water can be easily liberated by using the same method we discussed for generating hydrogen on the Moon: electrolysis.

Rocket fuel ingredient number two, oxygen, required to "burn" the methane, comes from either the water produced in the hydrogen/carbon dioxide reaction used to produce the methane fuel or from other, similar chemical processes. Recall that oxygen is present in water, carbon dioxide, and many other compounds found to exist on Mars.

It is important to note the significant potential impact of this approach to space exploration. The amount of fuel needed to return to Earth from Mars for a human expedition is significant. Making the fuel required for the return leg of the trip *in situ*, rather than lugging it with you, might be the difference between making the mission happen, or not. The launch mass savings, hence cost savings, are enough that mission planners are already taking notice and seriously considering this option for future missions.

JUPITER, SATURN, URANUS AND NEPTUNE: OUR NEXT STOPS

The gas giants of the solar system, Jupiter, Saturn, Uranus and Neptune, have very similar atmospheres. They are mostly composed of hydrogen, with varying abundances of helium, methane and other compounds mixed in. It is not too difficult to imagine a future with orbiting platforms positioned around these planets, extracting from them hydrogen and methane, thereby producing fuel for a succession of spacecraft destined for elsewhere in the solar system.

ASTEROIDS AND COMETS: ABUNDANT AND DISPERSED GAS STATIONS

Unlike the harbinger of battle found in the Bayeaux Tapestry, comets are much more akin to a neighborhood fueling station – waiting for vehicles to stop and fill up. The key is their water. Comets are nature's filling stations; created at the beginning of our solar system, they circle the Sun endlessly, and are available for any who have the ability to catch them and use their precious frozen water for drinking, radiation shielding, or fuel.

As we will use water found at the Moon or on Mars, we will similarly make use of the water frozen in comets. Because that is where we typically see them, we tend to think of comets as a resource for only the inner solar system. If so, we think incorrectly! Circling the solar system, and orbiting the Sun well beyond the orbit of Pluto, is a vast halo of comets, collectively known as the Oort Cloud. Named after its discoverer, Jan H. Oort, the cloud extends outward from the Sun to nearly 30 trillion kilometers—beyond half the distance to the nearest extrasolar system. Once we get our rocket ship beyond the planets of our solar system, this potential source of water might be available for journeys beyond.

Clearly, we are in the earliest stages of our use of chemical propulsion for deep-space exploration. The propulsion systems in use today are barely adequate for most of the missions we task them to perform. These in-space stages are ignited and burn most of their fuel within the first few hours, if not minutes, of launch. Following this is the very long coast through space until the spacecraft either ignites a set of engines to slow down and be captured into orbit around its destination or not. The "not" results in mission failure. We still lug all the fuel for the mission with us from the moment the rocket leaves the launch pad, paying the penalties imposed by the rocket equation on whatever missions we are flying.

But it does not have to be that way. The concepts, techniques and processes described in this chapter are not new, nor are they conceptually difficult. They will, however, require that we change our way of looking at space exploration missions. They will drive us to take the long view, planning for multiple missions and the establishment of an in-space transportation infrastructure. Until these "filling stations" are established on the Moon, Mars, a comet or an asteroid, the status quo will remain. And the status quo is clearly unacceptable.

17

HUMAN EXPLORATION

Space and Time! now I see it is true what I guess'd at,
What I guess'd when I loaf'd on the grass,
What I guess'd while I lay alone in my bed,
And again as I walk'd the beach under the paling stars of the morning.

Walt Whitman, from *Song of Myself*

IN 2004, United States President George Bush announced that American astronauts will be returning to the Moon before the year 2020, and then setting their sights upon the planet Mars. When this was announced, the advanced space technology community was energized. After all, it has been many years since Neil Armstrong first placed his footprint in the lunar dust. Surely a return to the Moon will use the latest technologies to provide an infrastructure for sustained lunar exploration and operations. Having gone to the Moon in the 1960s with 1940s rocket propulsion technology, one could imagine a 2015 human lunar return using solar- and nuclear-thermal rockets, solar-electric-propelled cargo tugs efficiently carrying cargo from low Earth orbit, and piloted vehicles aerocapturing back into Earth orbit for rendezvous with the International Space Station.

Unfortunately, the budget and schedule realities soon became apparent and the notion that new technologies would play a central role in the near-term return to the Moon faded. With international commitment to completion and utilization of the International Space Station and the need to do this with the existing space shuttle orbiter fleet requiring a significant fraction of the NASA budget, not much is left over for a lunar return. Add to this the inherent time required to develop, test, and implement new technologies (see Chapter 9, "Technological Readiness") and it becomes painfully clear that in order to meet the deadline of returning to the moon before 2020, there simply is not enough time to use much, if any, new technology. NASA had to begin designing flight hardware *now*—using mostly existing technologies—and innovation would have to wait.

America's vision for space exploration is larger than simply the human lunar return. Robotic scientific exploration of the solar system will continue and it is here that we will begin to learn how to use new technologies to live off the land in space, one day transferring that capability to the larger and more propulsive intense needs of human exploration. Many of the technologies described in this book will first be used on smaller, less-expensive, and certainly riskier robotic missions of exploration in deep space. Once they are proven and widely used by our robotic explorers, they will be considered mature enough to be considered with that most valuable of cargo—human beings.

Which technologies might be scalable from robotic missions to human? First some numbers. The European Small Missions for Advanced Research and Technology-1 (SMART-1), launched in 2003, was propelled to lunar orbit by a solar-electric propulsion system. The SMART-1 spacecraft weighed 367 kilograms and was launched on an Ariane-5 rocket. It was a one-way trip to the Moon. By contrast, the Apollo 10 mission launched in 1969 sent three astronauts into lunar orbit for two days and then returned them safely to the Earth. The Command and Service Modules together weighed over 30,000 kilograms and were launched on the Saturn V rocket. These are two missions that reached and orbited the same destination. However, the robotic mission did not have to consider a safe return to the Earth nor did it carry three astronauts! It also used a state-of-the-art solar-electric propulsion (SEP) system, using the free energy provided by the Sun to produce electrical power and drive the Hall thruster-based primary propulsion system. Clearly, the propulsion system required for a 30,000-kilogram human-piloted vehicle is different from that required for its smaller robotic kin. Fortunately, many technologies have a growth path that will some day enable them to be used for human exploration.

AEROCAPTURE

Oddly enough, the use of a planetary atmosphere as a brake, bleeding off excess kinetic energy from interplanetary travel, began with human exploration. Because it was never used, it is not well known that aerocapture (see Chapter 10) was the backup plan for the Apollo astronauts as they returned from the Moon. Should their primary chemical propulsion system have failed, they would have achieved Earth orbit using aerocapture. Fortunately for the Apollo Program, unfortunately for advocates of aerocapture technology, the maneuver was never executed and it remains undemonstrated to this day.

How might it be used? In the near term, it would be very useful for spacecraft returning to Earth from the Moon or Mars. Recall that most of the mass of an interplanetary spacecraft, human or robotic, is devoted to its chemical propulsion system. Trade studies show that replacing the chemical propulsion system with the structure required for aerocapture saves anywhere from 20 to 70% of the total propulsive mass. Its use on lunar missions might not be widespread, at least initially. The Moon is close and few operational scenarios call for astronauts to remain in Earth orbit on their return. More likely, they will aeroentry into the atmosphere, like Apollo, for an immediate landing. Mars, however, is a different situation.

There is vigorous scientific debate on the topic of planetary biological protection. If life still exists on Mars, might it be a risk to Earth if it was accidentally brought back here by one of our visiting expeditions? Might we be facing a scenario similar to that depicted in the movie, *The Andromeda Strain*? Bluntly, we do not know and we might not know until we actually send a crew there to explore and search for signs of life—past or present. Many have suggested that we isolate a retuning crew in Earth orbit for an extended period of time to make sure that no unwanted biological passengers accompanied them on their trip home. There might even be some sort of decontamination facility in low Earth orbit to examine any materials returning from Mars without endangering the Earth's biosphere. If this were to be the case, then the rationale for Earth orbit aerocapture becomes quite strong. Why carry propellant at the expense of other needed cargo or crew for a maneuver that can be accomplished using the resources that nature has already provided?

The other leg of Martian exploration that would benefit from aerocapture is at Mars itself. Various studies by NASA and others have shown the mission-level benefits of using a non-propulsive capture at Mars. The mass savings from its use ripple through the mission design and result in fewer initial rocket launches to get the Mars vehicle into space and lower costs per mission.

A gradual infusion of this technology into robotic exploration will be required before human mission planners seriously consider its use, primarily due to the perceived risk. After all, braking a 1,000-kilogram robotic spacecraft into Mars orbit is far different from a human-piloted spacecraft that weighs 600 times as much. The technological barriers are basically the same: thermal protection, precision terminal guidance (to make sure the spacecraft is where it needs to be), and knowledge of the target atmosphere. All of these potential barriers are actively being broken down for the smaller, robotic systems.

SOLAR-ELECTRIC PROPULSION

Sustained human expansion beyond Earth not only requires the transportation of the people, but also of the cargo (supplies) needed to sustain them. Since cargo is not as sensitive to the time spent in deep space, lower-thrust systems such as solar-electric propulsion (see Chapter 11) might be used to efficiently carry such supplies on the long voyages between planets. Taking advantage of plentiful sunlight, SEP-based vehicles might be the cargo ships of deep space. Why would one use SEP instead of chemical propulsion for this job? The short answer is: *efficiency*. Electric-propulsion systems are considered low thrust, which means that they cannot accelerate large masses to high speeds very quickly. However, given enough time, they can accelerate these masses to significantly higher speeds than chemical rockets and do so much more efficiently. In general, SEP systems are 10–100 times more efficient (per kilogram of fuel) than their chemical counterparts. Though they may take longer to attain high speeds, they do it efficiently—thus lowering the overall propellant requirements for the mission.

For example, consider the possibility of building a continuously occupied human base on the Moon. All elements of the base (the habitat, the vehicle that carries the crew from the Earth to the Moon and back again, the supplies, etc.) are launched into Earth orbit on board conventional chemical rockets. Before the humans depart for the Moon, the habitat and supplies are sent ahead from low Earth orbit using large solar-powered, solar-electric-propelled vehicles that take them to low lunar orbit. Once the habitat arrives at the Moon, the crew departs from Earth using a higher thrust chemical propulsion system. They descend to the surface, either in the habitat or separately, joining with it after landing. While this scenario could be accomplished without SEP, it would require more launches—just to carry the fuel that would not be required when electric propulsion is utilized.

In the far term, solar-electric propulsion might play a key role not only in transferring cargo from Earth to Mars, but also in transferring the crew. With its inherent efficiency being most advantageous on deep-space missions, very large solar-powered vehicles might one day be used to carry people. The idea of a "solar clipper" using solar arrays hundreds of meters long to power thrusters requiring hundreds of kilowatts of power is not new. Again, the benefit of using the technology comes down to the mass required to accomplish the mission. And less mass is always better. In the terrestrial real estate business, the mantra is "location, location, location." For space exploration, it is "propulsion, propulsion, propulsion." And solar-electric propulsion provides a much more efficient and economical approach than chemical rockets—it reduces the amount of fuel required, saving on launch cost and overall system mass.

It should be noted that not all solar-electric propulsion technologies are applicable to human exploration. The most widely used deep-space SEP system today is a gridded ion thruster. The physics of their design limits them to mostly lower-power, lower-mass applications—perfect for robotic exploration, but woefully inadequate to support humans. More likely, electromagnetic propulsion systems like those that use Hall and Magnetoplasmadynamic (MPD) thrusters will be used. They require much higher power, but are capable of higher thrust without losing too much of their efficiency.

SOLAR-THERMAL PROPULSION

Using the same plentiful sunlight that makes solar-electric and solar-sail propulsion possible, solar-thermal propulsion (see Chapter 12) offers a much higher thrust than either of them. This higher thrust is at the expense of efficiency, though it is still twice as efficient as its chemical propulsion brethren. Using sunlight to superheat propellant, thus achieving very high exhaust velocities, solar-thermal propulsion would appear to be a good candidate for moving large cargo in near-Earth space. The system-level benefits of the technology are somewhat different from those offered by solar-electric propulsion. The overall mission-level propellant requirements, hence the total mass to be launched by rocket, would be less than an all-chemical system. They would not be as low as that provided by SEP, however the thrust of a solar-thermal system would be significantly higher, making it better for moving large masses around fairly quickly.

Solar-thermal propulsion might be most useful when we achieve a level of interplanetary exploration beyond simply visiting the Moon or Mars. When humans are routinely moving around in the Earth–Moon system and beginning to mine the plentiful resources of the asteroids, we will see solar-thermal propulsion begin to play a key role. It can be used to quickly and efficiently carry the minerals and ore from asteroid mines back to the Earth for processing. Such processing facilities might be in Earth orbit or on the Moon. These solar-thermal propulsion systems will be high thrust, allowing them to move large payloads, with high efficiency, thus lowering their propulsion mass and cost. Since cruise time will not be as long as that required for deep-space missions to Mars and beyond, the continuous, low thrust of a SEP system will not be as attractive as solar-thermal propulsion for these applications. Solar-thermal propulsion offers an attractive balance of quick trip times and lower mass—essential factors for a future space-based industry delivering real products from space for paying customers.

The technologies required to field a solar-thermal propulsion system are closely aligned with those that will be required for some in-space manufacturing applications. For example, large, inflatable solar concentrators used to heat the propellant might be adapted to use on the Moon or asteroids for heating rock as part of the smelting process. Just as our ancestors learned to obtain copper from ore by heating it beyond its melting point, so will we learn to emulate the process in deep space as the foundation of a non-terrestrial industrial base. These "simple" manufacturing processes will have to be relearned in this radically different environment.

The fuel for the solar-thermal propulsion system will initially be launched from Earth with the propulsion hardware. At first, this will also be true for the oxidizer required for the smelting operation. However, economics will almost certainly drive us to derive them from the resources of space. Lunar regolith contains oxygen (as well as many elements and minerals, many of which will be very useful for other manufactured goods)—required not only for our human crews to breathe but also for smelting. We can extract the oxygen to enable our factory workers to use it by adding hydrogen and heating it—using the solar concentrators developed for the solar-thermal propulsion system. On the question of where the hydrogen comes from, recall that hydrogen is an excellent propellant for solar-thermal propulsion systems and is abundant in the form of water ice in comets. Hydrogen can be obtained from water by passing a current through it in a process called electrolysis. (Which also releases the other elemental component of water: *oxygen!*)

Data from the Lunar Prospector mission suggests that water ice may be present on the Moon near its poles, protected from the heat of the Sun by being in the perpetual shadows created by lunar craters. The water or ice need only be resupplied infrequently as the smelting process does not consume the hydrogen, but merely uses it as a catalyst. A closed system can recycle most of the oxygen and hydrogen many times, producing excess oxygen from the regolith in each cycle. There are undoubtedly more efficient or creative methods for extracting the minerals needed for an industrial society from lunar or asteroid resources, and the expertise of creative chemical and industrial engineers will most certainly be essential.

SOLAR SAILS

Unfortunately, the limits of modern engineering and materials science will limit the use of solar sails (see Chapter 13) to small robotic spacecraft and cargo until some sort of breakthrough occurs, or until we learn how to construct ultra-thin, large sails in space from *in situ* resources. Recall that solar sails derive their propulsion by reflecting sunlight. Lots of sunlight, falling on a very large and ultra-lightweight structure is required to achieve even the most modest of robotic science missions. (A 150-meter square solar sail with an areal density of less than 15 grams per square meter is required to propel a 400-kilogram spacecraft!) Typical human-support sized craft will weigh 40,000 kilograms or more. Clearly, much larger and lighter weight sailcraft will be needed to accelerate these ships. (Don't forget, Newton's law clearly applies—for a given force, the higher the mass, the lower the acceleration. As the spacecraft mass goes up, the acceleration provided by the propulsion system, in this case a solar sail, goes down.)

Robert Frisbee of NASA's Jet Propulsion Laboratory has looked at the problem and estimated the sizes of sails needed to support human exploration missions to Mars. A Martian solar-sail cargo vehicle capable of sending metric tons of supplies to astronauts exploring the planet Mars would be 2 kilometers on a side (assuming a square sail configuration) and require approximately 4 to 5 years to make the one-way journey. This assumes a sail using first-generation technologies to achieve an areal density of approximately 13 grams per square meter. While the overall mass density of the sail is comparable to those that will be fielded in the next several years, the overall size of the sailcraft (hence the issues that will be encountered in simply manufacturing the sail itself) places this application at least 15–20 years into the future.

SPACE TETHERS

Electrodynamic tether propulsion, using the free resources of sunlight and the Earth's magnetic field and ionosphere, could be implemented almost immediately with significant benefit to human explorers. As described in Chapter 15, fairly modest length tethers could be fielded in support of the International Space Station to provide reliable, low-cost reboost and orbit-maintenance propulsion. They can be used on virtually any system that is required to be in low Earth orbit for an extended period of time—from the massive International Space Station, to the more modest orbiting hotels proposed by entrepreneurs such as Bigelow Aerospace.

Momentum Exchange Electrodynamic Reboost tether systems may take much longer to achieve than their strictly electrodynamic cousins, but they are no less revolutionary. A network of MXER tethers surrounding the Earth, the Moon, and eventually Mars, may provide a reliable, low-cost and continuously operating transportation system for a robust and expanding human civilization. By initially implementing the system in the Earth–Moon system, human explorers will have the ability to get supplies, materials, and finished products to and from the lunar surface and Earth orbit—regularly, and at low cost.

The challenges to human exploration are significantly greater than those facing the precursor robotic explorers due simply to the complex and massive infrastructure needed to sustain human life in space. The abundant resources of space, fortuitously provided by Mother Nature, appear to offer us ways to reduce the burden we have assumed by carrying all our supplies with us as we leave the home planet. Unfortunately, we have not yet developed the technologies required to harness these resources to the level that they can be readily used without significant risk. Unless we begin to try, however, they will never be available! Any viable, long-term and sustainable plan for human exploration must begin to use solar-system resources or is surely doomed to failure.

18

DEFENDING THE EARTH

Nor the comet that came unannounced out of the north flaring in heaven,
Nor the strange huge meteor-procession dazzling and clear shooting
  over our heads,
(A moment, a moment long it sail'd its balls of unearthly light
  over our heads,
Then departed, dropt in the night, and was gone;)

Walt Whitman, from *Year of Meteors*

NEAR misses have happened many times in Earth's long history. A fragment of asteroid or an errant comet approaches our world closely, skips through the upper atmosphere and back into space leaving a noiseless contrail, or bathes the evening sky in the shimmer of light reflected from its tail. But sometimes, a near miss or deflection shot morphs into a direct hit.

In the beginning, four and a half billion years ago, such impacts were a good thing. In those days, comet swarms criss-crossed the infant solar system. When they impacted a planet, lots of water, methane, and ammonia were released. Earth's oceans and atmospheric ingredients, necessary for the formation of life, are thought to have been deposited in this manner. It's not impossible that early life was actually transferred from Mars to Earth in the aftermath of an impact on the low-gravity Red Planet.

As the solar system evolved, potential impactors of the inner planets became rarer. And instead of being beneficial for terrestrial life forms, the impacts that occurred were destroyers rather than creators.

The fossil record reveals that many mass extinctions disrupted the terrestrial biosphere during the last billion years or so. The most famous of these, which occurred about 65 million years ago, ended the reign of the dinosaurs. Many believe that the impact of a celestial body in the 10-kilometer range is at least partially responsible for the fall of the dinosaurs and the rise of the mammals.

No humans existed on Earth at the time of the Great Fall, but we can imagine how the thunder lizards or our crude, early mammal ancestors might have reacted as the huge space mountain descended toward ground zero, in what is now the Yucatan.

Perhaps they had been aware of subtle shifts in skylight for a few days prior to impact, if the object was a comet. Perhaps the normal cycles of sleep and the hunt had been disrupted by these strange changes in the sky.

If the meteorite was instead of asteroidal origin, the first warning would have been flames in the sky and sonic booms. Perhaps some creatures craned their necks toward the intruder, perhaps others cowered in fear. It made no difference.

The Cretaceous–Tertiary (K/T) meteorite struck the Earth with enormous kinetic energy. To get some idea of the effects on Earth's environment, consider that the largest hydrogen bombs in the Cold War arsenals of the super powers had about 1,000 times the explosive yield of the weapons that destroyed Hiroshima and Nagasaki in 1945. The energy released in the Yucatan by the K/T impact had something like half-a-million times the yield of the largest H-bombs.

Such a large explosion, in one place and time, was certainly not a good thing for Earth's ecology. A huge fireball would have instantly extinguished all or most life within thousands of kilometers of ground zero. An enormous mushroom cloud would have risen to the stratosphere and ultimately dispersed to enshroud the entire planet. Firestorms would dart across the landscape; giant tsunamis would haunt the oceans and crash against coastal regions. It's not impossible that other seismic events—volcanoes and earthquakes—would be triggered globally by the impact.

Soon, Earth temperatures would decline as stratospheric dust blocked the Sun. Vegetation would die back, followed by the huge herbivores. Deprived of their food source, carnivores at the top of the food chain would soon follow them into extinction.

Among major land animals, only primitive mammals (who probably could hibernate in their burrows and sleep away the bad times) and flying, feathered dinosaurs—the ancestors of today's birds—survived the end of the Cretaceous era. Ultimately, their descendants would evolve and radiate to fill ecological niches left by the demise of the larger creatures.

Certainly, climate change, volcanism, and perhaps other factors contributed to the dinosaurs' extinction. But the enormous, rude visitor from space certainly helped to push them along.

Fine, you say, but all this occurred 65 million years ago. In all that time, the solar system has been cleared of at least some of the asteroids and comets that might threaten the Earth.

True, but lots of dangerous space rocks still exist in solar space. From time-to-time, they still whack the Earth. The last impact of consequence occurred in 1908, in Tunguska, a sparsely populated part of Siberia. With a diameter of about 100 meters, the Tunguska object stuck the Earth with the force of a large hydrogen bomb. If its trajectory had been slightly different, casualties would be numbered in the millions.

In 1998, John Remo published an inventory of near-Earth objects (NEOs) capable of some day impacting the Earth. Based on telescopic surveys, he estimates that there are about 20 that could cause mass-extinction events—those with diameters in excess of 5 kilometers. There are approximately 400 in the 2-kilometer size range, about 6,000 that are approximately 0.5 kilometer in size, and about 100,000 0.1-kilometer Tunguska-sized city killers.

Based on these numbers, which do not reflect random comet storms from the Oort Cloud, we can expect to lose a city once every few centuries and experience continent-scale devastation every couple of millennia. At million-year intervals, impacts with global consequences will occur and mass extinctions will be experienced at intervals of tens of millions of years.

Our planet seems to whirl through a cosmic shooting gallery. Is there anything we can do?

NUCLEAR IMPACT-THREAT MITIGATION

One approach to mitigating the threat of impending NEO impacts is the application of nuclear "devices." For obvious reasons, this is the method favored by Hollywood special-effects specialists. If a NEO or errant comet is detected on an Earth-colliding trajectory, this scenario proposes that a fleet of nuclear-tipped space vehicles should be dispatched to either blow the intruder to smithereens or alter its trajectory.

Although nukes could be applied as last-ditch Earth defense, exclusive reliance on this "nuclear option" has a significant drawback. This problem has a lot to do with our ignorance regarding the mechanical properties of small celestial objects.

As discussed by Carl Sagan and Ann Druyan in *Comet*, there is a class of comets in orbits so elliptical that they graze the Sun at perihelion. Astronomers have noted that some of these "Sungrazers" that survive a close solar pass actually split or "calve" into two or more separate objects after perihelion.

But comets can calve under much less extreme conditions. In 1994, Comet Shoemaker–Levy 9 fragmented into many smaller bodies after a close pass of Jupiter. These objects were later observed to impact the giant planet.

A nuclear blast can hardly be considered a precision device. It is not impossible that nuclear devices activated near an Earth-intercepting comet (or near-Earth asteroid of cometary origin) might split the comet into two, or more, highly radioactive fragments each targeting the Earth, rather than altering the comet's trajectory. So alternatives to nuclear explosives should certainly be considered in any dedicated Earth-defense program. Again, we'll look toward the resources provided by Mother Nature to see if any can be applied to the problem.

USING SPACE RESOURCES TO MITIGATE THE THREAT

Nukes may still serve as last-ditch NEO-interceptors in spite of the calving problem. But if we have impact warning times measured in decades and can very accurately compute asteroid and comet trajectories, a number of solar options exist that could turn a direct hit into a near miss.

One of these is the mass driver. A mass driver is essentially a solar-powered electromagnetic catapult that could be mounted on the asteroid's or comet's surface. Parcels of material could be flung at high velocity into space (possibly to be intercepted later for use in space manufacturing). By Newton's Third Law (every action has an equal and opposite reaction) the offending object's trajectory would be altered by this process.

But this approach requires either despinning the NEO or carefully timing mass-driver ejections. If the mass is thrown overboard while the NEO is spinning, the total change in velocity, at least in the desired direction, will be much harder to achieve. Another approach without these disadvantages is to mount a solar-sail shroud around the offending celestial object. This would have the effect of increasing both NEO area and reflectivity. Solar radiation pressure would alter the NEO's solar orbit, rendering it more elliptical. If warning time is sufficient, a trajectory targeting the Earth could be morphed into a near miss by simply changing the amount of sunlight reflected off the body.

In 2005, a NASA probe called Deep Impact deployed a high-velocity penetrator aimed at Comet Temple 1. The resulting jet of volatiles from the comet's nucleus (Figure 18.1) offers some other deflection possibilities.

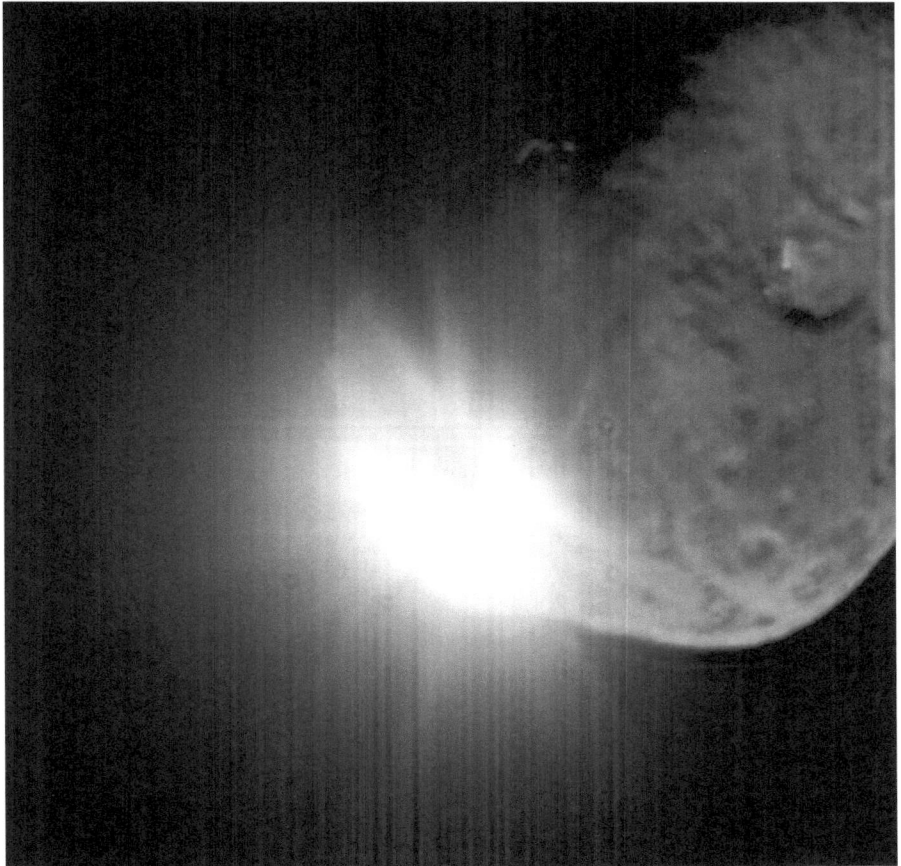

FIGURE 18.1 The nucleus of Comet Temple 1, 13 seconds after the high-velocity arrival of the Deep Impact Probe in July 2005. (Courtesy NASA)

One might take advantage of this approach by maintaining a solar concentrator such as the one utilized in solar-thermal propulsion (see Chapter 12) in position near a volatile-rich, Earth-threatening NEO or comet. The resulting jet of material, heated by the solar concentrator, would thrust the NEO or comet in the opposite direction.

Another possibility is high-speed impacts with the threatening object. Kinetic energy and linear momentum transfer would alter the trajectory of the NEO or comet, or possibly calve it into fragments too small to survive passage through the Earth's atmosphere.

With enough advanced warning, the trajectory of an Earth-threatening NEO could be altered by changing its reflectivity. As was discussed in the chapter on solar sails, we know that sunlight, while having no mass, does have momentum. This momentum can be transferred to the NEO, slowly pushing it and changing its trajectory, by either being absorbed by the NEO, or used twice as efficiently by reflecting from it. Using sunlight pressure to ever so slowly adjust its orbit until the NEO no longer poses a threat to the Earth makes the many issues surrounding the use of nuclear weapons in space superfluous. To change its reflectivity, and thus its trajectory, the NEO could be coated with highly reflective chalk dust or wrapped in a large solar sail intentionally designed to serve that purpose at the end of its flight. What Mother Nature throws our way can be gently moved OUR WAY using her own indigenous resources!

FURTHER READING

Life-threatening meteorite impacts are discussed in many popular sources. Notable among these is *Comet* (Random House, New York, 1985), which was authored by Carl Sagan and Ann Druyan. A more recent source is *Origins* (Norton, New York), which was written by Neil DeGrasse Tyson and Donald Goldsmith in 2004.

Inventories of near-Earth objects and possible impact-threat mitigation approaches are also reviewed in a number of references. One is Gregory L. Matloff's *Deep-Space Probes*, 2nd edn (Springer–Praxis, Chichester, UK, 2005).

The calving of Comet Shoemaker-Levy 9 during its first Jupiter pass and the subsequent 1994 impacts on Jupiter were well-observed and photographed astronomical events. One source of further information regarding these events is Eric Chaisson's and Steve McMillan's *Astronomy Today*, 3rd edn (Prentice-Hall, Upper Saddle River, NJ, 1999).

For more information regarding mass drivers and similar devices, consult the space-resource literature. The classic reference in this field is Gerard K. O'Neill's *The High Frontier* (Morrow, New York, 1977).

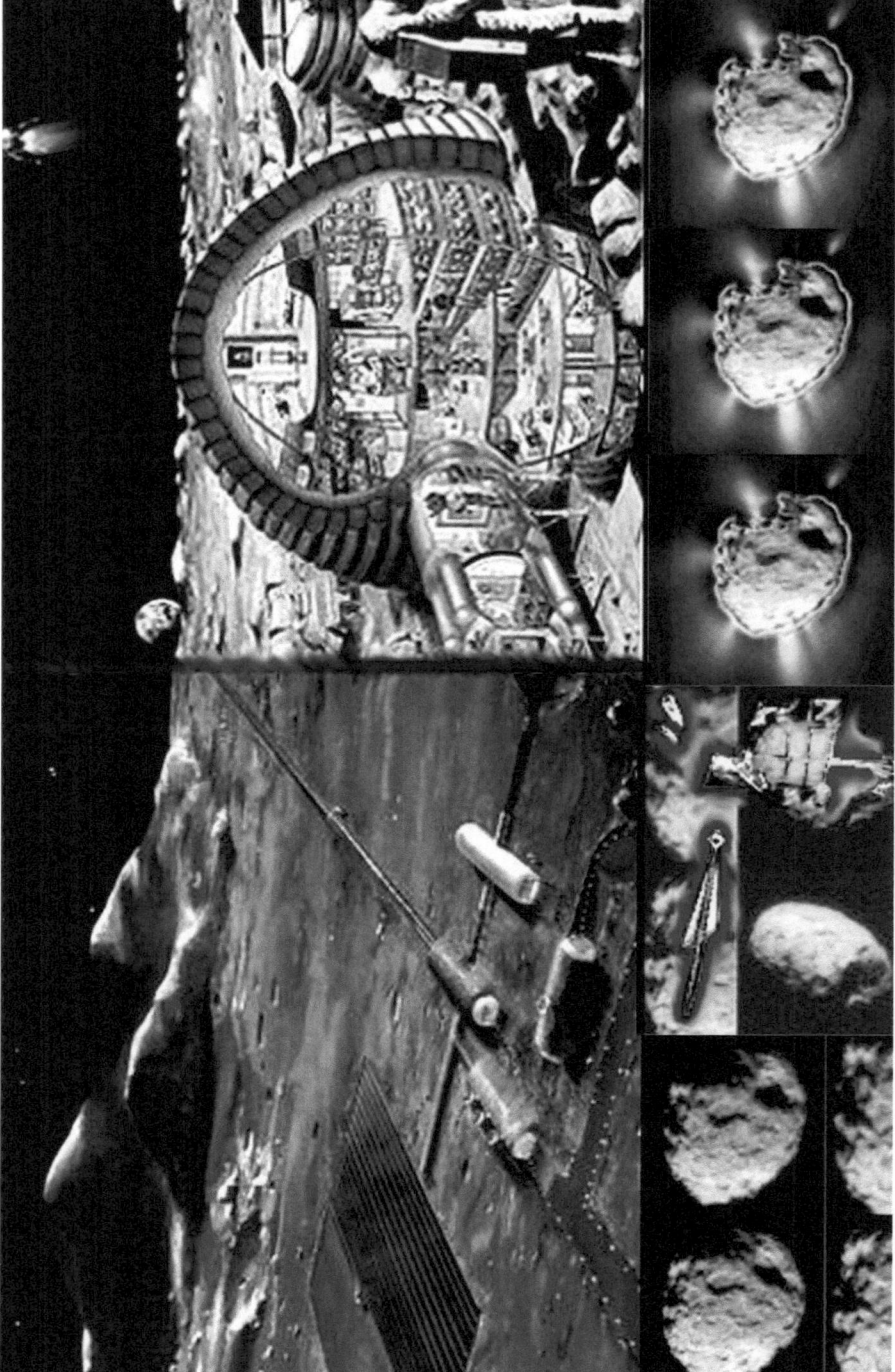

19

SPACE MINERS

Mightier than Egypt's tombs,
Fairer than Grecia's, Roma's temples,
Prouder than Milan's statued, spired cathedral,
More picturesque than Rhenish castle-keeps,
We plan even now to raise, beyond them all,
Thy great cathedral sacred industry, no tomb,
A keep for life for practical invention.

Walt Whitman, from *Song of the Exposition*

CIVILIZATION began with the birth of metallurgy and the first application of the mine to tap the hidden wealth of Planet Earth. Since the Egyptian Bronze Age, every temple and cathedral, every castle, every public building has depended upon stone from the quarries and tools constructed from mined metal.

The United States would be a less-developed country today if it were not for the fortuitous 1849 discovery of gold in California. The resulting westward migration and development of transcontinental trade routes dependent upon clipper ships, wagon trains, and the Transcontinental Railroad united the North American continent. In a similar fashion, mining solar-system resources may well result in a codependent unity of space habitats—a kind of "Federation of Sol."

Science fiction abounds with space prospectors and the space pirates who rob them. One author who has constructed an elaborate future history around these concepts is Larry Niven. He has described the exploits of his intrepid asteroid-mining "belters" in a number of short stories.

One aspect that would-be space miners must deal with is the uneven distribution of the vast resources of the solar system. Other issues are the vast distances and long travel times separating a "space mine" and a space-manufacturing center near Earth. Some of the technologies described in previous chapters can greatly aid the exploitation of solar-system resources.

POSSIBLE NEAR-EARTH RESOURCE LOCATIONS

The most obvious near-Earth location to place our hypothetical space mine is the Moon. After all, it is relatively close and round trips can take only about one week. The Moon has no atmosphere to interfere with solar energy flux required to power mining equipment, and no ocean or biosphere to pollute. So why not simply place our mining equipment on the lunar surface and go to work?

There are several reasons why the Moon is not an ideal location for a large-scale mining facility. First, the lunar surface is at the bottom of a gravity well—the lunar escape velocity is about 2.4 kilometers per second. This means that a substantial infrastructure must be built up simply to loft resources from the lunar surface to a space-manufacturing facility in free space.

The fact that the Moon lacks oceans or other surface water deposits is another problem for lunar mine planners. Unless water can be gathered and transported from possible cometary deposits near the lunar poles, volatile water-derived rocket fuels used to blast astronauts or cargoes off the lunar surface would have to be transported from Earth, rather than mined from lunar soils. One way to resolve this would be to build a solar-powered catapult—a "mass driver"—to loft mined material from the lunar surface without the use of chemical rockets.

The apparent rarity of *in situ* lunar water poses another problem for those planning to live off the lunar landscape. Unless they are living near a polar subsurface water deposit, astronauts supervising robotic lunar mines will have to be extremely careful to recycle every drop of water that passes through their artificial biospheres. The importation of large quantities of water from Earth could economically doom the venture.

Another problem with the Moon's suitability as a self-sufficient space mine is energy. Except for a few near-polar locations bathed in perpetual sunlight (or constant darkness), the Moon experiences 14 days of sunlight followed by 14 days of darkness. Solar energy is problematic and nuclear fission will almost certainly not be independent of expensive terrestrial intervention.

A final issue with which would-be lunar miners must contend is the availability of valuable lunar resources. A good deal of media attention has been devoted to the possibility of mining significant quantities of a very special isotope of helium, known as helium-3, from the lunar regolith. Most of the helium found on Earth is composed of 2 protons and 2 neutrons. The Sun produces another type of helium, one with 2 protons and 1 neutron. This is helium-3.

Helium-3 is indeed a very valuable resource. Very rare on Earth but common in the solar wind, this low-mass helium isotope could be burned with deuterium (hydrogen with an extra neutron) in our first-generation thermonuclear-fusion reactors. Unlike most possible fusion-fuel combinations, helium-3 and deuterium are clean burning. Very little radioactive waste would be produced by a fusion reactor that is fueled with deuterium and helium-3.

However, lunar regolith samples drawn by Apollo astronauts were only drawn in a few locations and never more than a meter or so below the surface. Although solar-wind-deposited helium-3 is present in some of these samples at very low concentrations, it would be unwise to plan a mining operation solely around this resource without much more extensive explorations of the lunar environment.

The Moon is a good place to get our feet wet and to try out new techniques. If a scientific lunar base fails, evacuation to Earth requires a trip of only a few days. It is a good place to locate observatories and other scientific facilities, but it is an imperfect location for a space mine.

Perhaps we should not be to hasty in downgrading lunar mining prospects, however. There is one possibility of constructing a profitable lunar-mining enterprise that makes use of lunar resources and does not require lofting processed material from the lunar surface into space.

Using lunar resources, it should be possible to construct large quantities of photovoltaic cells that would convert sunlight directly into electricity. In time, a large fraction of the Earth-facing lunar surface could be paved with these cells. Microwave transmitters with football-stadium sized apertures could beam the energy toward Earth.

Even when one takes into account the various inefficiencies—for example, the lunar day/night cycle, the fact that sunlight will generally strike the Moon's surface obliquely, sunlight-to-electricity conversion efficiencies are generally less than 20%, some microwave energy will be absorbed by Earth's atmosphere, etc.—it is evident that most or all of Earth's energy could be supplied by a solar-cell-paved Moon.

But is this an economically attractive alternative? The cost of electricity in New York City is currently about $0.2 per kilowatt hour, or about $1 per 20 million joules. If a lunar solar facility can supply 15% of Earth's electrical energy per year (about 10,000 billion watts for 30 million seconds), the producer could easily earn $150 billion per year. Whether or not this is a profitable venture will depend upon the cost of construction, maintenance, and the time-value of the money involved.

Yes, space mining will be expensive, but projected incomes like these may impress a few wealthy entrepreneurs.

Another relatively near-space resource possibility are the near-Earth objects (NEOs). As discussed in Chapter 18, it would be a good idea for our emerging global civilization to rearrange the orbits of some of these asteroids or extinct comets that might ultimately impact the Earth. Why not mine them at the same time?

One interesting class of NEOs are of the Aten class. These celestial bodies (comets or asteroids) have semi-major axes less than 1 AU and aphelia greater than 0.983 AU.

NEOs with low eccentricity, low inclination, and perihelia and aphelia close to 1 AU are those most likely to collide with the Earth or Moon within the next 10,000 years unless they perform Earth–Moon gravity-assist maneuvers to be flung into other regions of the solar system. Because of their orbital parameters, they will be relatively easy for astronauts to visit—with relatively short mission duration and facilitating expeditions to the Moon and Mars.

To see if any of the objects pose a significant threat, one must determine their approximate sizes. Members of this class with diameters much less than 50 meters would probably disintegrate during entry into Earth's atmosphere and therefore should not be considered as major threats. Also, the payoff from mining very small remote objects such as these is likely to be negligible.

Asteroid 2001 CQ36 seems to be representative of the NEOs that may pose a threat while at the same time holding promise of economic return from being mined. With an estimated diameter between 90 and 200 meters, this NEO circles the Sun once every 332 days. It approaches the Earth closely in 2021, 2022, 2031 and 2033, with its 2031 minimum

Earth-separation of only 0.02 AU (about 3 million kilometers—less than 10 times the Earth–Moon separation).

If one assumes a 150-meter diameter for this object, a spherical shape and a specific gravity of 2 (intermediate between water and rock), then this object has a mass of approximately 3 billion kilograms. If this asteroid were to impact the Earth and 50% of its mass reached the surface, the resulting explosion would be about equivalent to that of a 30-megaton thermo-nuclear explosive, the largest of these monster weapons in the Cold War arsenals of the US and USSR. Depending upon the impact location, tens of millions of people could die during such an event and property damage could be immense.

Asteroid 2001 CQ36 has an orbital eccentricity of 0.176 and an inclination of 1.3 degrees. Its perihelion and aphelion are 0.774 AU and 1.11 AU, respectively.

PRELIMINARY EXPLORATION

As with any solar-system body, robots should precede humans to the Atens. A robotic probe with a mass of a few hundred kilograms and propelled by solar-electric propulsion could rendezvous with Asteroid 2001 CQ36 after a flight of only a few months. The craft could survey the object remotely, determining its dimensions and rotation rate. If a soft landing is made, the asteroid's surface composition could be studied.

Before a full-scale mining expedition is launched, astronauts should also visit the NEO. A round trip to Asteroid 2001 CQ36 would probably take about a year, much less time than a peopled mission to Mars. Astronauts could conduct seismic studies and return rock samples to Earth for detailed analysis. The mass budget for such a mission using chemical propulsion would be a few hundred thousand kilograms.

MINING THE ATENS

Many of the concepts relating to asteroid mining were first seriously studied during the 1970s as part the Princeton/NASA–Ames investiga-tions of space habitation and manufacturing. Any asteroid-mining venture will actually be a three-step process.

The initial mission phase concerns the transfer of equipment and personnel to the celestial object to be exploited. Then, the equipment must be installed upon the surface of the asteroid and the asteroid's rotation must be stopped. Finally, as much of the asteroid material as possible must be returned to a space manufacturing facility, most likely in Earth–Moon space.

The basic tool to be used in asteroid mining is most likely be the mass driver. Described by a number of authors, including Gerard K. O'Neill and Brian O'Leary, the mass driver is essentially a solar-powered electromagnetic catapult. It can be mounted on the lunar surface and used to fling payloads of mined material into space to be grappled by a mass catcher—a space equivalent of a catcher's mitt. Alternatively, it could be operated as an in-space propulsion system expelling any convenient material, such as ground-up, spent rocket stages or it could be mounted on an asteroid and utilized to propel the asteroid through space by throwing away packets of asteroid rock.

The exhaust velocity of a mass driver used as an in-space propulsion system would be as high as 8–10 kilometers per second; it could accelerate as much as a million kilograms per year. Its mass would be in the vicinity of 100,000 kilograms and its length would be in the multi-kilometer range. For use on a celestial body's surface, the mass driver could be optimized for higher mass throughput and lower exit velocity. Much of the technology of these devices has been tested and is shared with terrestrial devices including magnetically levitated (maglev) trains.

A mining expedition to an Aten object might look something like this:

Possibly under its own power, the robot mass driver departs Earth orbit first, possibly utilizing a low-energy trajectory (such as those discussed in Chapter 4) and lunar gravity assist, to effect rendezvous with the NEO. The human crew follows on a higher energy, chemical-rocket-propelled trajectory to minimize exposure to solar and galactic radiation.

Arriving at the destination, astronauts first set up camp on the surface of the NEO, using NEO material as cosmic-radiation shielding. The mass driver is then utilized to cancel the rotation of the NEO—a necessary, but not very major task since most asteroids rotate with periods of many hours.

The mass driver is then used as a rocket, the exhaust being parcels of asteroid material flung into space. The NEO's trajectory is gradually altered so that it is captured as a distant satellite of the Earth, where workers from an orbital space manufacturing facility can disassemble it. As the entire process may take years, the astronauts on the NEO might be rotated back to Earth and replaced by fresh crews several times.

A tremendous amount of material could be returned to earth orbit for use in constructing solar-power satellites, radiation shields, and for other applications that we can only imagine. If only one-third of the estimated mass of Asteroid 2001 CQ36 could be returned in this manner, about one-billion kilograms of material would be available for space manufacturing.

TAPPING MORE DISTANT SPACE MINES

As humanity's in-space infrastructure develops, it will become possible to economically tap resources ever-deeper in the solar system. Main-belt asteroids, small planetary satellites, and the atmospheres of giant planets are but a few examples of the destinations and resources available there.

Using solar concentrators such as those developed for solar-thermal propulsion (see Chapter 12), sunlight could be used to separate space resources according to melting point. Huge solar sails constructed from *in situ* materials could be utilized to transfer mined material back to near-Earth processing facilities.

As discussed by John S. Lewis, robotic space miners could ultimately tap the atmosphere of Uranus, shipping home large quantities of helium-3 for use in terrestrial thermonuclear reactors, or perhaps interstellar spacecraft.

In the farther future even the Kuiper Belt Objects and Oort Cloud comets could provide resources for our growing civilization. Space mining could support a very large solar-system population at a high standard of living. And it may serve as the springboard to the stars.

FURTHER READING

An excellent reference dealing with solar-system resources and how to access them is John S. Lewis, *Mining The Sky* (Addison-Wesley, Reading, MA, 1996). To learn about properties of Atens and other NEO classes, consult Kathrina Lodders and Bruce Fegley Jr, *The Planetary Scientist's Companion* (Oxford University Press, New York, 1998).

Mass-driver theory and experiments are described in a number of sources. Check out, for example, papers by Gerard K. O'Neill and Brian O'Leary (and others) in *Space Manufacturing Facilities II: Proceedings of the Third Princeton/AIAA Conference*, ed. Jerry Grey, American Institute of Aeronautics and Astronautics, New York, 1977. Other sources include Gerard K. O'Neill, *The High Frontier* (Morrow, New York, 1976) and Brian O'Leary, *The Fertile Stars* (Everest House, New York, 1981).

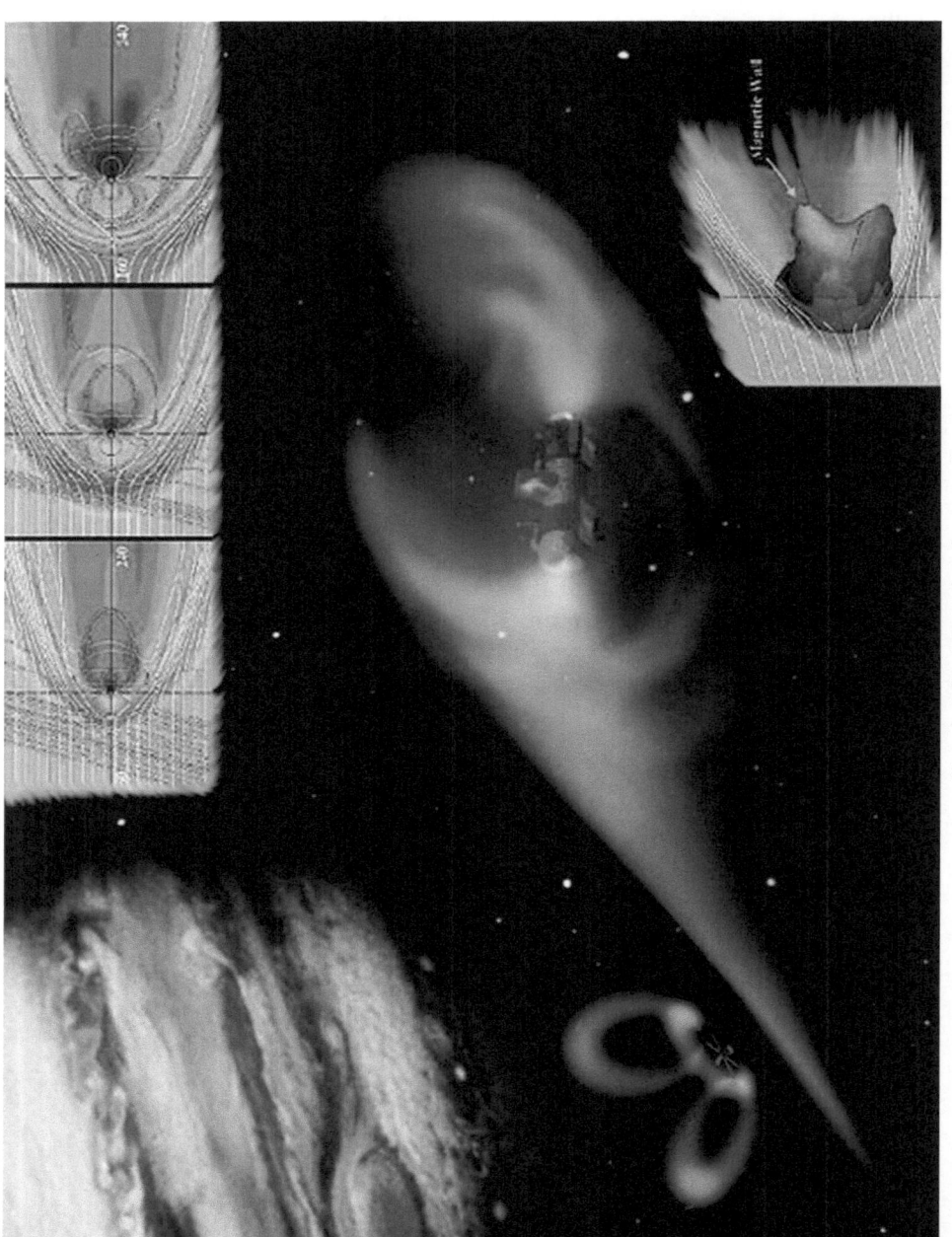

Magnetic Wall

[See also Plate VI in the color section]

20

SOME EXOTIC POSSIBILITIES

To see a world in a grain of sand
And a heaven in a wild flower,
Hold Infinity in the palm of your hand
And Eternity in an hour.

William Blake, from *Auguries of Innocence*

SO what's new? Many of the ideas expressed in this book are not new. It is only now, however, after decades of fundamental research and laboratory experimentation that many of these technologies are at the point where they can be seriously considered for real applications in space. For the foreseeable future, space exploration technologies will be based on ideas that are as old as the Space Age itself, if not older. There were 60 years of laboratory research between Hermann Oberth's discussion of electric propulsion in 1939 and the 1998 flight of an ion engine as the primary propulsion system for the Deep Space 1 mission. It took 90 years for Konstantin Tsiolkovsky's solar sail to go from an

219

interesting idea to a realizable propulsion system. As we move closer to fielding systems such as these, it is rightful to ask if there are any great new ideas in the pipeline that might be available for the *next* generation of space explorers.

Not so very long ago, interplanetary travel by rocket was a fantasy, considered by most to be unworthy of serious consideration. People who publicly speculated about life beyond the confines of Earth's biosphere risked scorn, ostracism or worse.

We have indeed come a long way. But it is worthy of note that the approaches presented here to interplanetary travel and development rest heavily upon early twentieth-century physics and mid-twentieth-century technological projections. It may be a good idea to follow the poet William Blake and attempt to "Hold Infinity in the palm of your hand" and speculate about possible future developments that may ease the road ahead.

Many of the suggested exotic approaches to space flight that we are about to consider may prove to be physically impossible or impractical to implement. But it must be remembered that only one need succeed. Even if they all fail, the seed of a realistic breakthrough concept may be planted by this discussion in the fertile soil of a reader's mind.

PLASMA SAILS

Around 2000, the space propulsion community was abuzz with excitement about a newly proposed propulsion system that offered the potential of travel to the outer edges of the solar system riding the solar wind—and consuming very little propellant – called Minimagnetospheric Plasma Propulsion, or M2P2. The brainchild of Dr Robert Winglee of The University of Washington, the M2P2 would take advantage of the fact that the Sun is constantly shedding high-speed charged particles (called the solar wind) which race outward from it at speeds near 400 kilometers per second.

The M2P2 would work by creating around a spacecraft a miniature version of the Earth's ionosphere, taking advantage of the fact that like-charge electrically charged particles in a plasma repel each other. The ionosphere is the part of Earth's atmosphere that is ionized by solar radiation, producing a relatively dense plasma environment. A plasma is nothing more than a mixture of positively and negatively charged particles that are typically too energetic to recombine and form neutral molecules. The plasma is mostly "trapped" on the Earth's magnetic field lines due to electromagnetic interactions. As a general rule, charged particles moving in a magnetic field will experience a force due to their motion through the field. The forces acting on the ionospheric plasma tend to send it spiraling along the field lines, moving between the Earth's north and south magnetic poles. The solar wind, which is also a plasma, often interacts with the Earth's ionosphere, causing it to be depressed on the sunward side and extended on the opposite side. The depression on the sunward side is actually exerting a propulsive force on the Earth, though at a scale so small compared with the mass of the Earth as to be insignificant.

The M2P2 concept takes advantage of the forces acting on the Earth, in miniature, to propel a spacecraft sporting its own ionosphere. An M2P2 propulsion system would have on board a very strong magnet and a plasma source. The field generated by the magnet would trap the plasma and create a multi-kilometer-wide plasma bubble around the spacecraft. The solar wind would then hit the bubble at enormous speeds, with like-charged particles repelling each other, theoretically causing the bubble-encased spacecraft to be blown outward like a balloon tossed into the air on a windy March day. Since the solar wind does not decrease in intensity nearly as rapidly as sunlight, the forces acting on the plasma sail should theoretically be able to accelerate it for much longer periods, allowing the craft to attain very high speeds. Dr Winglee calculated that a plasma sail craft might be able to send a probe to Jupiter in only 1.5 years compared to the 5 years required for a conventional, chemically powered spacecraft.

As with any new technology, there remain many unanswered questions about the ultimate viability of the M2P2 as a spacecraft propulsion system.

- Will the plasma bubble remain intact after encountering the much more energetic solar wind? Or will it be ripped away from the spacecraft, leaving it motionless as the wind blows by?

- Can the very powerful magnets required to make it work actually be built and launched affordably? (Magnets can be very heavy and therefore difficult to launch.)

- What is the plasma leakage rate from the bubble? If too much plasma leaks away, then the overall efficiency of the system drops and might soon make it no more attractive than state-of-the-art electric propulsion systems.

More research (and time) are needed to answer these questions.

MAGNETIC PROPULSION

Every child has played with magnets and learned about their attraction and repulsion. Since the Earth, Jupiter, and the Sun are enormous cosmic magnets, it is not surprising that mature researchers have devoted serious effort to investigating the feasibility of magnetic space propulsion.

Much of the modern work in this field germinated from a paper by Gregory Matloff and Alphonsus J. Fennelly in the September 1974 issue of *JBIS*. This paper addressed the problem of collecting interstellar ions for the interstellar ramjet (see Chapter 6). A superconducting magnetic solenoid was proposed as an interstellar ion-collection device.

In research by Robert Zubrin and Dana Andrews, reviewed in Gregory Matloff's *Deep-Space Probes*, plasma-physics computer codes were employed to examine the trajectories of interstellar ions encountering the solenoid's magnetic field. It was found, however, that most proposed magnetic interstellar-ramjet ion-collection devices actually function as ion reflectors, not ion collectors.

Although bad for ramjet fans, these superconducting ion reflectors (dubbed "magsails") turned out to have interstellar applications. They can theoretically function as very effective drag brakes, to slow a speeding starship down to planetary velocities by reflecting sufficiently large quantities of ionized gas in the local space, wherever that might be. Large magsail field radii (measured in hundreds or thousands of kilometers) and long deceleration durations (measured in decades) are necessary to accomplish this task for a large, fast starship.

Magsail derivatives were considered that could theoretically reflect interplanetary ions and operate as a propellantless-propulsion device between the planets. It appeared that direct interaction with Earth's magnetic field, allowing for magnetic-assisted liftoff from Earth's surface, might be possible.

But alas, these proposals proved to be a little overoptimistic. In research reviewed in Gregory Matloff's *Deep-Space Probes*, Giovanni Vulpetti and Mauro Pecchioli demonstrated that thermal effects in the inner solar system would limit the operation of superconducting magsails.

ANTIMATTER PROPULSION

Antimatter is real. Particle physicists have been creating antimatter for at least half a century as a byproduct of high-energy particle accelerator research. But what is it? First theorized by Paul Dirac in 1928, antimatter—particles with the same mass as their normal matter cousins but with opposite charge—is regularly produced in many facilities used for particle physics research. The antimatter counterpart of the electron is the positron. The proton's antiparticle is the antiproton, having the same mass as a proton but an opposite charge. When a particle and its antiparticle meet, they do something very interesting and potentially very useful: they annihilate each other and convert all their combined mass into energy. It is the most efficient mass-to-energy conversion process known—far more efficient than nuclear fission or fusion.

Created when atoms of normal matter smash into each other at relativistic speeds, antimatter particles quickly encounter normal matter atoms and annihilate them. These events have been studied and recorded for decades. Even under hard vacuum conditions, atoms of antimatter will eventually cross paths with atoms of normal matter and annihilate them. It is thus very difficult to store antimatter for any significant length of time. Conventional vacuum chambers cannot be used to store these elusive atoms because they too are made of normal matter and readily annihilate when the antimatter atoms come into contact with their walls. It should be noted that nature, too, produces antimatter. When high-energy cosmic rays (which are nothing more than atoms moving at very high speeds with an origin in deep space) enter the Earth's atmosphere, they collide with atoms in the atmosphere. Some of the enormous energy of the cosmic rays is converted into matter–antimatter pairs, which soon interact and annihilate.

Since antimatter atoms have essentially the same physical properties as normal matter, they follow the same laws of physics. Charged particles in the presence of a magnetic field experience a force acting upon them that is perpendicular to both the lines of magnetic force and their direction of motion. Similarly, charged particles in the presence of an electric field will also experience a force, this time aligned along the electric field line. Any charged particle, matter or antimatter, will experience these forces—though, since their charges are of opposite sign, the force will push the antimatter particle in a direction opposite to the normal matter particle. No matter, the bottom line is that electric and magnetic fields can be used to "trap" antimatter particles in vacuum, thus mostly preventing them from coming into contact with the vacuum chamber walls. The antimatter ions spiral along the magnetic field lines, reflecting back from both the north and south poles of the magnet in much the same way that ions are trapped along the Earth's magnetic field lines in the ionosphere. These magnetic traps are called Penning traps and have been successfully used to trap charged particles for extended periods of time.

How, then, might these energetic atoms be used for space travel? And how is this relevant to "living off the land in space"? To answer the first question, matter–antimatter collisions, under controlled conditions, can be used to create enormous amounts of energy. It is, in effect, a very efficient battery. Unfortunately, it is also a very expensive battery. Global production of antimatter hovers near a nanogram (that's 0.000000001 gram) per year. For a viable propulsion system, one would need several grams of antimatter. And, by the way, it would have to be stored with nearly 100% efficiency lest it should inadvertently become a very powerful bomb, which would be the result if a significant fraction of the antimatter were to get free and annihilate all at once. If you recall that this reaction is orders of magnitude more energy efficient than nuclear fission or fusion, you can then imagine the disaster that would ensue as a result of such an event. For comparison, the total energy released by the space shuttle engines (and its 1.7 million kilograms of fuel) during launch is the equivalent of 0.1 gram of antimatter! Unfortunately, antimatter is also one of the most expensive items on earth to produce. In 1999 Dr George Schmidt of NASA calculated the cost to produce 1 gram of antimatter at 62.5 trillion (yes, trillion) dollars. How, then, might this be relevant to "living off the land in space"?

Given its highly volatile nature, the large-scale production of antimatter would be a very dangerous venture. Ideally, it should be produced in a remote location so that any industrial accident would not kill a large nearby population and devastate a continent. It should be produced in a location where the energy required to manufacture it is plentiful and cheap—one doesn't get all this energy for nothing. It still takes a lot more energy to create antimatter than can be extracted from it. The beauty of the antimatter is its high-energy density and storability for use on relatively small spacecraft. One place where these conditions are simultaneously met is the Moon. As sunlight is plentiful, large solar array farms could be assembled to provide the power required to run the accelerators and make the antimatter. And the Moon certainly satisfies the "remoteness" criteria should there be an accident, endangering only the astronauts/workers and not an unwilling population. In this case, nature provides us with the real estate and the power—now all we humans need to do is discover how to affordably mass-produce the stuff.

BREAKTHROUGH PROPULSION PHYSICS

Barring some revolutionary change in the known laws of physics, nature seems to have placed a nearly insurmountable barrier between us and the rest of the universe—the immense distances that must be traversed for any explorers, robotic or human. Crossing many light years in a human lifetime—or even in the lifetime of a human civilization let alone the potential lifetime of the spacecraft making the journey—appears to many to be nothing more than a dream. Even using antimatter-driven spacecraft, or the solar-sail-propelled version described in Chapter 13, it will take centuries, if not millennia, to reach nearby stars. Unless nature has provided an as-yet unknown "escape clause," we may never journey beyond the closest stellar systems.

Nevertheless, new theories and phenomena are being reported in the scientific literature that at least give spacefarers pause for thought—and hope. Perhaps science-fictional concepts like wormholes, space-warping drives, and vacuum energy are both real and usable? If so, will our "live off the land" philosophy extend to journeys beyond Sol? The answer is a resounding, "yes!" If, and only if such exotic theories are validated will we see significant human exploration beyond the home solar system—we simply cannot otherwise take enough energy or propellant with us on such a journey. Aside from limited and infrequent interstellar journeys using sails, there seems to be no possible method of establishing the galactic empire of science fiction books and films.

From 1996 through 2002, NASA funded the Breakthrough Propulsion Physics (BPP) Project at the NASA Glenn Research Center. Though the funding was modest, the project attempted to rigorously investigate many of these exotic physics ideas and recent theoretical or experimental claims regarding phenomena that might lead to faster-than-light travel. Using the peer-review process and involving credible members of the scientific community, the BPP Project investigated quantum tunneling, vacuum energy extraction, Woodward Transient Inertia, and others. The Project also assessed similar theories and work sponsored elsewhere. No obvious breakthroughs were found. Does that mean that nature has not provided humanity with a way to reach the stars? No—it does mean, however, that we don't know if she has or not.

The following concepts may have a Technological Readiness Level near Zero in AD 2006. But who knows what they might lead to in the far future?

REPLACING THE ROCKET: ANTIGRAVITY

No one can deny that the launch of a large rocket is an awe-inspiring and dramatic sight. It is also dangerous and inefficient. Space travel would be safer and less expensive if a gentler device could be developed to replace the reaction engine, something perhaps akin to the magic carpets of ancient fables.

One conceptual rocket replacement—an approach that has captured the hearts of both science fiction authors and Hollywood special-effects experts—is antigravity. Simply strap on your spacecraft, plug in your convenient antigravity device, and float slowly up into the cosmos.

As discussed by Gregory Matloff in *Deep-Space Probes*, there is a theoretical basis for antigravity. In the early instants of the universe, gravity was linked with the other forces of nature, including electromagnetism. Might an electromagnetic machine be possible that could convert Earth's gravitational attraction into repulsion?

Some theoretical work suggests that the answer to this question is "yes." A number of experiments, some using test masses suspended over rapidly spinning superconductors, have been performed.

To date, these experiments have not proven that the theory is valid. And any experimentally suggested antigravity effects are very small, perhaps non-existent or just very close to the limits imposed by experimental precision. But antigravity remains a tantalizing concept and research will certainly continue.

REPLACING THE ROCKET: THRUST MACHINES

Imagine a machine that could directly convert electrical or mechanical energy into spacecraft kinetic energy at high efficiency. Such a device could act in a similar manner to an antigravity device, to accelerate a spacecraft, aircraft, flying car, or flying city skyward with little or no reactive thrust. Humans, like birds, would become creatures truly at home in three dimensions.

From time to time, such devices have been proposed and tested. To date, all have proven to be impossible to implement or have been shown to directly violate the basic laws of mechanics or thermodynamics.

The classic space drive of this type, the "Dean Drive" of the 1960s, is discussed by Eugene Mallove and Gregory Matloff in *The Starflight Handbook*. Awarded a US patent, this device purports to convert rotary motion into linear motion. But controlled laboratory tests on similar devices demonstrated that they will not function as advertised.

A more recent device, developed by British engineer Roger Shawyer, is still under evaluation. This device purports to function by converting direct-current electrical power in a microwave tube directly into thrust. To date, no one has independently demonstrated the device.

TAPPING ZPE AND GETTING SOMETHING FOR NOTHING

The ultimate free lunch is the universe itself. The vacuum is not truly empty; in fact, at the smallest scales it is quite dynamic with lots of energy and mass blinking into and out of existence in the twinkling of a quantum's eye. But something happened about 14 billion years ago to stabilize one of these quantum fluctuations. Space, time, energy, and ultimately matter erupted in what we call the Big Bang. It then expanded and ultimately coalesced into the familiar environment we find around us. Now, if only we could do the same thing and tap this energy of the vacuum, or "zero-point energy" (ZPE), on even a small scale, we would have the energy at our disposal to sail right up to the speed of light!

It is quite possible that the inertia of all objects with mass is due to interaction with this quantum foam. Perhaps we could learn some day to "polarize" the vacuum so that a starship might accelerate off in a selected direction like a massless particle traveling at speeds that we can only imagine.

Sadly, all experimental efforts to date to tap ZPE have been fruitless, which leaves them at a Technological Readiness Level of Zero.

HYPERSPACE SHORTCUTS

As discussed in Chapter 6, space–time warps seem even less likely today than ZPE machines. But research in these fields should continue in spite of what seems to be a low probability of success. Enormous amounts of energy—actually universal amounts—may be required to produce a wormhole. And exotic, purely theoretical energy fields must be manifested from the quantum vacuum to maintain them.

Breakthroughs have occurred before—much of twentieth-century physics "broke through" the nineteenth-century belief that Newton and Maxwell would eventually describe how everything works and came about. They were wrong, and perhaps today's physicists are also wrong. If so, then the development of either a warp drive or a ZPE machine would have enormous consequences for humanity. We should be skeptical, follow the scientific process and, above all, keep our inquisitive minds open to new and different ideas.

Let us suppose that we find a way for nature to provide us with a revolutionary space drive. What then might we want to accomplish?

INSTANT EARTHS

If we wish to terraform a marginally habitable world—such as Mars—and make it more amenable to terrestrial life, we are taking on a long-term and expensive project. We first must use devices such as solar collectors or reflectors to increase (in the case of frigid Mars) solar radiant flux. Then, we must add lots of water and atmosphere—probably by impacting volatile-rich bodies (comets) from the Asteroid Belt or Kuiper Belt. Next we mix well and stir—probably for a few centuries—to distribute the ocean and air as desired. And finally, as our final ingredient, we add life of increasing complexity.

If future humans find such worlds in neighboring solar systems and travel to them on voyages requiring many centuries, imagine the frustration of residents in the "worldships" as they endlessly orbit and peer down upon the slowly evolving planet. They will certainly wonder whether any method exists to speed up the process.

Hollywood special effects make terraforming look easy. In one of the *Star Trek* movies, a "genesis" device is used to almost instantly alter the character of a non-living world and equip it with a functioning biosphere.

Such a device seems impossible today, but it is unreasonable to limit the frontiers of nanotechnology and genetic engineering. Even if it proves impossible to rapidly create a biosphere on a marginally habitable world, it may ultimately be possible to reengineer higher terrestrial organisms so that they can instead adapt to the environments of Mars-like worlds.

THE RED-EYE SPECIAL

Space is a very big place and foreseeable propulsion technology has its limitations. Trips to Mars may always require months; flights to more distant planets, Kuiper Belt or Oort Cloud Objects may always require decades; and trips to nearby stars may require centuries.

One suggested way to relieve the tedium of a long interplanetary or interstellar voyage, retard aging, and reduce the strain on your craft's life-support system, is to take the Red-Eye Special and sleep your way through space.

Such a technique—called suspended animation—is an old standby for the science fiction author. In Stanley Kubrik's screen adaptation of Arthur C. Clarke's *2001, A Space Odyssey*, most of the crew sleep their way between Earth and Jupiter.

As well as reducing life-support requirements and possibly extending lifespan, hibernation might increase tolerance to cosmic radiation, thereby reducing shielding requirements on long space flights.

Humans do not hibernate. So is there any real hope that we can suspend the animation of a human crew during a long-duration space voyage?

Although humans can't yet hibernate, the proteins responsible for hibernation in certain rodents have been isolated. One significant aspect of possible hibernation in humans, long overlooked, is the psychological effects of this process. Recent research addresses possible psychological and behavioral consequences of induced human hibernation.

At one time, it was fashionable to assume that ultracold storage of recently deceased organisms could be accomplished using cryogenic techniques. Although meat preservation can certainly be accomplished in this manner, it seems most unlikely that deep-frozen brains can ever be revived.

FURTHER READING

Terraforming has attracted a good deal of attention in recent years. One excellent sourcebook describing "conventional" terraforming techniques, authored by M. J. Fogg, is *Terraforming: Engineering Planetary Environments* (Society of Automotive Engineers, Warrendale, PA, 1995).

The new science of the very small (i.e. nanotechnology) is developing rapidly. A standard reference in this field is K. Eric Drexler's *Engines of Creation: The Coming Era of Nanotechnology* (Anchor/Doubleday, New York, 1986).

For a review of human hibernation concepts prior to 2000, consult Gregory L. Matloff's *Deep-Space Probes*, 2nd edn (Springer–Praxis, Chichester, UK, 2005). If you'd like to check out more recent work in this field, consult papers included in *Proceedings of the Fourth IAA Symposium on Realistic Near-Term Advanced Scientific Space Missions: Missions to the Outer Solar System and Beyond*, ed. G. Genta, Aosta, Italy, July 2005 (Politecnico di Torino, Turin, Italy, 2005).

For a discussion on thrust machines, such as the "Dean Drive," see Eugene Mallove's and Gregory Matloff's *The Starflight Handbook* (Wiley, New York, 1989).

You can read about early tests on Shawyer's microwave engine in his paper "The Development of a Microwave Engine for Spacecraft Propulsion," *JBIS Space Chronicles*, vol. 58, Supplement 1, pp. 26–31 (2005).

Many popular references treat various aspects of zero-point energy and how we might learn to exploit it. One very readable source is Henning Genz's *Nothingness: The Science of Empty Space* (Perseus, Reading, MA, 1999). Attempts to apply this energy source to spacecraft propulsion are reviewed in Gregory Matloff's *Deep-Space Probes*, as are speculations regarding warp drives.

Advanced Chemical Propulsion

Today's chemical rockets are almost at the theoretical limits of their performance. Researchers are seeking to extend these limits by improving storage life, developing lightweight propellant management systems, and increasing the capability of cryogenic propellant systems to support deep space exploration; and by developing new propellants.

Aerocapture

Aerocapture describes a method of using a planet's atmosphere to assist in decelerating an incoming spacecraft in order to achieve orbit, without the need for on-board propellant. Friction between the spacecraft and the planetary atmosphere slows the spacecraft, allowing it to enter orbit.

Electric Propulsion

Electric propulsion systems use electromagnetic, electrostatic, or electrothermal energy, instead of chemical energy, to accelerate propellant and achieve low, but more efficient, thrust for longer periods of time. The primary benefits of electric propulsion are decreased launch mass, increased delivered mass, and reduced mission time as compared to chemical systems.

Solar Sail Propulsion

Thin, lightweight reflective solar sails produce thrust by harnessing the "pressure" of sunlight. Because they use no propellant, solar sails can open new regions of the solar system for exploration and long-duration observation.

Emerging Propulsion Technologies

The In-Space Propulsion Technology Program actively identifies and evaluates innovative candidate technologies for development. One such technology is the Momentum-eXchange/Electrodynamic-Reboost (MXER) tether, which may provide a reusable in-space infrastructure for high-thrust and high-efficiency spacecraft propulsion. A MXER tether facility would be capable of boosting spacecraft from Low Earth Orbit to higher energy orbits, and could dramatically enhance the performance of all other space transportation technologies while reducing launch costs.

[See also Plate VII in the color section]

21

SIGHTS ON CENTAURUS

O to sail to sea in a ship!
To leave this steady unendurable land,
To leave the tiresome sameness of the streets, the sidewalks and the houses,
To leave you O you solid motionless land, and entering a ship,
To sail and sail and sail!

Walt Whitman, from *A Song of Joys*

SOME day there will be deep-space ships that we can enter and live within, ships that can support us on millennial journeys and sail the dark seas of the interstellar void. And even if we utilize the resources of Sol's system wisely, husbanding rather than blindly consuming, we will some day need such ships. For the Sun's lifetime is finite.

The Sun that nurtures all terrestrial life is near the midpoint of its 10-billion-year life as a hydrogen-fusing main-sequence star. As it ages, its luminous output increases. In other words, the Sun is slowly getting brighter.

To compensate for this slow roasting, humans or their descendants may ultimately construct large sunscreens from asteroids or comets and use these devices to reduce the solar flux. But no matter, in about 1.5 billion years, the oceans will start to boil; and the atmosphere will begin to escape into space.

For those who have elected to remain as groundlings, emigration will be the only non-fatal alternative. Colony ships may shuttle populations between dying Earth and suddenly clement Mars. Ultimately, even Mars will become to hot and these far-off sons and daughters of Adam might seek out new homes farther from the Sun—on worlds such as Europa or Titan.

But even this will be a half measure. When the Sun finally leaves the main sequence, it will expand to become a giant star, enveloping or roasting all the inner planets. Within 100 million years or so as a giant, the Sun will have exhausted all its thermonuclear fuel sources and will shrink once again, ending its career as a dim, white dwarf star.

No matter what we do, the Sun's giant phase will be an eviction notice. If we have not already done so, it will be time to leave our familiar solar lands behind and venture forth onto the galactic seas.

Given both the difficulties of interstellar travel and the lure of stellar worlds, it is difficult to predict when interstellar exploration will begin. It may all depend upon what is learned about the hypothetical planets that may attend the Alpha Centauri suns (see Chapter 2).

If the next generation of space telescopes discover one or more Earthlike worlds circling Alpha Centauri A or B, robotic star probes at least are a near certainty within the next century or two. On the other hand, if these two stars are attended only by Marslike worlds or comets and asteroids, the pace of interstellar exploration may be slower. Will the public want to fund expensive expeditions to distant, non-living worlds, when plenty of these exist within our own solar system? Even though we cannot predict the timing of the first interstellar expeditions from Earth, we have a fair idea of the technologies they may use, barring breakthroughs like those discussed in Chapter 20.

THE FIRST STARSHIPS

Of the various technologies reviewed in Chapter 6, light sailing seems to be the best current choice of systems to propel humanity's first starships. Ramjets are infeasible, antimatter is difficult to store and too expensive, and the socio-political problems inherent in nuclear-pulse propelled spacecraft may be insurmountable. Only the light sail remains.

As discussed by Ben Finney and Eric Jones, there are two basic approaches to interstellar expansion via light sail. In keeping with previous terrestrial folk migrations, humans might launch a series of relatively slow but comparatively inexpensive solar-sail starships, which might be called "slow boats" or "nomads." If large-scale and stable government funding is available, we might instead select to travel in more expensive but speedier "fast ships" propelled by the radiation pressure of a solar-powered laser, beaming energy from the inner solar system to the distant starship.

At intervals of 100,000 years or so, random stellar motions bring stars closer to the Sun than the 4.3 light-years separating us and the Alpha Centauri system. It is also possible that a subluminous star closer to us than the Centauri suns will be discovered. In the discussion that follows, however, it will be assumed that the destination for our first robotic and peopled starships will be Alpha Centauri.

THE SLOW BOAT TO THE STARS

Even though it will be less costly than a fast ship, construction of an interstellar slow boat will be a monumental engineering task. To allow our robot or human population the luxury of a 1,000-year transit time, solar-sail engineering will be stressed to its physical limits.

At their best, current sails could complete this journey in about 7,000 years. It will be necessary to develop very thin, strong and temperature-tolerant sails that are joined to the payload by diamond-strength cables. (Unless, that is, the payload consists of thin-film electronics deposited on the anti-sunward sail face, as suggested by Dr Robert Forward in his "Starwisp" concept.)

The 0.2-AU perihelion distances considered for current-technology extrasolar sails will also not be adequate. To take full advantage of solar radiation pressure, our starship must approach the sun closer than 0.1 AU. Care must be taken that solar radiation pressure does not blow the craft prematurely from the inner solar system during its inbound, pre-perihelion trajectory leg.

Ultimate solar sails will most likely be manufactured in space, rather than launched from Earth. Sail manufacturers might first mine appropriate sail materials—perhaps aluminum, beryllium, scandium, niobium, or tungsten—from asteroids or comets rich in such materials. The nanometer-thin-film sail could then be manufactured *in situ* using a process called vapor-phase deposition. To prevent solar-radiation back pressure on the sail during the solar pass, the sail could be partially furled behind a chunk of machined asteroid rock or a multi-sail configuration could be utilized so that the sail would initially be "hove-to" in respect to sunlight.

If we are launching a small robotic starprobe, the unfurled sail dimension would be perhaps 1–10 kilometers. Starships capable of maintaining a human population for centuries will require sails 100–1,000 kilometers or larger in linear dimension. These ships could be small versions of the O'Neill space habitats discussed in Chapter 7.

Carbon could also be mined from comet nuclei for this venture. Space manufacturing facilities would utilize solar energy to create diamond-strength filamentary cables from this material, and the cables would be used to join the sail to the payload.

When the construction project is complete, the probe would be activated or the human population would board the habitat. The ship, perhaps still shielded from the Sun by its chunk of asteroid rock, would depart toward perihelion, perhaps utilizing a Jupiter-flyby and auxiliary solar-electric propulsion for pre-perihelion maneuvers.

If human occupants are on board, they may choose to sleep within solar-radiation shelters during the hours of the close solar pass. Alternatively, like the alien astronauts in Buzz Aldrin's and John Barnes' science-fiction novel *Encounter with Tiber* (Warner, New York, 1996), they may view the process from their acceleration couches, entranced by the spectacle of a star at close range.

As the ship departs perihelion, the sail is slowly unfurled and ballast is released to maintain moderate acceleration. By the time the ship has passed Jupiter, it may be moving at 1,000 kilometers per second. It is time to furl the sail and prepare for the long interstellar cruise.

One advantage of the photon sail as an interstellar propulsion device is its versatility. After acceleration, sail (and cables) could be wrapped around the habitat to produce extra shielding from cosmic radiation. Since Alpha Centauri A and B are both Sunlike stars, the sail could be unfurled again at journey's end for deceleration.

If humans or their descendents wait until the Sun's giant phase before launching colony ships to nearby stars, the enhanced solar radiant flux will result in shorter trip times. But a magsail or some such interstellar drag brake (see Chapter 20) would be necessary to decelerate from the higher interstellar cruise velocity. Such a technique would also be required if our colony ship is directed toward a subluminous red dwarf star, rather than a near-twin of the Sun.

Traveling between the stars somewhat limits our options for onboard power. A nuclear reactor is possible, as is an electrodynamic tether that would obtain power from the interstellar magnetic field, at the expense of a small reduction in the ship's kinetic energy.

Arriving in the destination solar system, the starship occupants could construct larger habitats from local asteroids or comets. Alternatively, they could land directly on Earthlike worlds or terraform planets like Mars. Some have suggested that human starfarers might become perpetual nomads, picking up roots almost immediately and proceeding toward another stellar destination. Perhaps this process would continue until they reached the home system of another intelligence. One can only speculate on the consequences of such a contact.

If human interstellar explorers encounter intelligent extraterrestrials, we hope that everyone gets along. But since interstellar humans would be interlopers in someone else's home solar system, perhaps our explorers will need a protocol to continue exploring until an uninhabited world is located.

FAST SHIPS

Slow boats require little assistance from their home world after launch. But what if the home population of a starfaring society allocates a high priority to interstellar exploration. Extensions of light-sail technologies that may become feasible in the future might decrease interstellar transit times to one or two centuries. But a massive in-space infrastructure would be required—one that would operate seamlessly for decades or centuries.

A fast ship would resemble a slow boat, with one very significant difference. Instead of being propelled by the radiation pressure of solar photons impacting against the sail, the motive force for a fast ship would be a solar-powered energy beam. Conceptually at least, the fast ship could overcome the limitations imposed by the inverse square law that governs solar radiant flux and therefore reach a higher interstellar cruise velocity than a slow boat.

Although a lot of progress has been made lately in demonstrating that the link between energy beam and sail can be maintained over short distances, maintaining this link over trillion-kilometer interstellar distances is far from easy. It will be necessary to either light-levitate the beam-power station in a stationary position between the Sun and the starship for decades, or have the power station follow the starship out of the solar system on a slower trajectory.

In either case, the beam must be maintained on the sail for the decades-long acceleration run. The slightest deviation and the mission will be lost.

Assuming that the engineering challenges can be overcome, there are three options for the power beam. We might consider the most efficient approach (from a momentum-transfer point of view) and project a neutral charged-particle beam against a ship-mounted magsail. Unfortunately, since high-energy particle beams are a component of missile defense systems, it may be difficult to engineer them and convince the world that their use would be for peaceful purposes only.

As an alternative, we might consider a high-power, collimated microwave beam, or maser. Although masers do not suffer from military secrecy requirements, and microwave technology is comparatively inexpensive, maser (and laser) beams are much less efficient than particle beams in transferring momentum to the starship.

A more significant issue affecting microwave beams is beam spread. Unless a corrective optical system is placed between the beam and the starship, the beam will spread too rapidly to be of use for accelerating even very large sails. And the application of such an optical system would only complicate the problem of maintaining the beam on the sail.

The third and currently favored alternative is to utilize a huge solar-pumped space laser operating in the ultraviolet, visual, or near-infrared spectral regions. Beam spread would be less of a problem here, but laser-technology is orders of magnitude more expensive than microwaves. And lasers do not solve the problem of maintaining the beam on the distant sail for decades.

There is no fundamental reason why clever engineers cannot ultimately solve these problems. But interstellar beamed-energy sailing cannot be said to currently have a high Technological Readiness Level.

SPREADING THROUGH THE GALAXY

In the long run, it makes no difference whether we engage in millennial interstellar voyages on board slow boats or centuries-duration voyages on board fast ships. Human colonists arriving in an alien solar system must practice the skills learned close to home and rapidly learn to live off the lands on which they settle.

One wonders how quickly human civilization, or any extraterrestrial intelligence, can spread across the Milky Way Galaxy. If we use slow boats, the rate of interstellar colonization will be slow. Most stars are subluminous red dwarfs rather than solar-type stars like Alpha Centauri A and B. Colonies established around red-dwarf stars will have little incentive to expand to other stars because of the trillion-year lifetimes of such stars. Another limitation on the expansion rate from colonies in such systems is the low stellar radiant flux, which will greatly reduce the interstellar cruise velocity of the slow boats, and not help the prospects of fast ships.

Addressing these problems is a task for future generations. It is enough that the technologies we are beginning to apply today promise to open at least the nearest galactic frontier for our remote descendants.

FURTHER READING

A classic work discussing both the technology and sociology of star travel is *Interstellar Migration and the Human Experience*, edited by B.R. Finney and E.M. Jones (University of California Press, Berkeley, CA, 1985). A recent review of interstellar-lightsailing techniques (including Starwisp) is G.L. Matloff's *Deep-Space Probes*, 2nd edn (Springer–Praxis, Chichester, UK, 2005).

Future close encounters between our Sun and other stars were predicted by a NASA team investigating the extrasolar trajectories of Voyagers 1 and 2. The results of this study are reviewed by Eugene Mallove and Gregory Matloff in *The Starflight Handbook* (Wiley, New York, 1989).

A number of sources have examined the sociological and technological problems of maintaining human life on board multi-generation starships. A recent reference is *Interstellar Travel and Multi-Generation Space Ships*, edited by Y. Kondo, F.C. Bruhweiler, J. Moore and C. Sheffield (Apogee Books, Ontario, Canada, 2003).

AFTERWORD

Astronomy and astronautics are dynamic and rapidly evolving fields. No matter what the intentions of the authors and editors, the lengthy process involved in the creation of a book like this one will render it partially obsolete by the time of publication. The purpose of this Afterword section, which is written in September, 2006, is to partially compensate for this unavoidable problem.

One fascinating new development is the arguments regarding the status of Pluto, which was classified as the solar system's ninth planet when writing commenced. The International Astronomical Union has now ruled that this world should be classified as a "dwarf planet," rather than a major solar-system member. Its orbit is more highly inclined and elliptical than the orbits of the major members of our solar system. And it is much smaller. In fact, it is no longer even the largest known member of the Kuiper Belt.

One very intriguing result in planetary exploration is a discovery by the NASA Cassini probe, currently orbiting Saturn. Saturn's tiny satellite Enceladus may have a sub-surface soup rich in organic materials, including the precursors of life. Much to the surprise of Cassini mission scientists, a plume of water vapor was detected rising from this tiny world.

But the team conducting Japan's ambitious Hayabusa mission to asteroid Itokawa may have been less fortunate. The probe may not have succeeded in its sample-collection effort. We won't know the answer to this one until the Hayabusa's return to Earth in 2007.

INDEX